BEN HAMMERSLEY is a British writer and public speaker, specializing in explaining complex technological and sociological topics to lay audiences, and is a high-level advisor on these matters to governments and business. He is the British Prime Minister's Ambassador to TechCity, London's Internet Sector; Innovator in Residence at the Centre for Creative and Social Technologies at Goldsmiths, University of London; a non-resident fellow of the 21st Century Defense Initiative at the Brookings Institution in Washington D.C.; Senior Fellow at the Royal College of Defence Studies, London; a member of the European Commission High Level Group on Media Freedom; a fellow of the European Policy Centre in Brussels, and Contributing Editor at the U.K. edition of WIRED magazine. He presents the BBC Radio 4 series, FourThought, and speaks in an average of 20 countries per year.

He is a fellow of the Royal Society for the Arts, the Royal Geographical Society, and the Royal Institution, and a member of Chatham House. He was awarded a UN Fellowship in 2011, and made a member of the Royal Institution channel Scientific Advisory Group in the same year. He is Ambassador for the National Media Museum, and a member of the International Academy of Digital Arts and Sciences.

APPROACHING THE FUTURE
64 THINGS YOU NEED TO KNOW NOW FOR THEN

BEN HAMMERSLEY

SOFT SKULL PRESS
New York, NY

APPROACHING THE FUTURE:
64 THINGS YOU NEED TO KNOW NOW FOR THEN

First published as 64 THINGS YOU NEED TO KNOW NOW FOR THEN
in Great Britain in 2012 by Hodder & Stoughton, An Hachette UK company
Published as NOW FOR THEN: HOW TO FACE THE DIGITAL
FUTURE WITHOUT FEAR in paperback in 2013

Library of Congress Cataloging-in-Publication is available
ISBN 978-1-59376-514-9

SOFT SKULL PRESS
New York, NY
www.softskull.com

Cover design by BriarMade
Printed in the United States of America

For Mischa

CONTENTS

INTRODUCTION

Writing a book about the future is, in most ways, futile. We can't possibly construct a narrative that will be true. The world is already too weird. But what we can do is show some of the dominant ideas that are shaping the future, and our present, and from those gain an understanding of the direction we're travelling in. That is what I have tried to do in this book. The 64 ideas are all interrelated and are, I believe, changing how we live, work, and relate to each other in ways that are completely new. Understanding them is the first and best step to dealing with our collective future. Each of the 64 is an ingredient, which added to another can make something delicious, or potentially very nasty. As we move solidly through the second decade of the twenty-first century, we do well to pay attention to these forces as they shape our lives. Thank you for reading, and please be in touch.

This was a hard book to write, and many people must be thanked. Rupert Lancaster was patient beyond the call, for which

I am immensely thankful. Helen Coyle was and is magnificent at getting the words out of me, and she and Tara Gladden are responsible for anything that sounds smart. The rest is my fault. I cannot thank them enough. Many thanks too to Kate Miles, Emma Knight, Alice Laurent for the magnificent cover, and everyone else at Hodder & Stoughton for their much appreciated efforts.

The researches of Hannah Whittingham, James Berrill and Saya Robinson were of supreme help, as were my conversations and improvisations with the thousands of people in front of whom I have lectured or opined at in the twenty-eight countries I've visited since I started work on this. Books like this aren't ever the work of one person. Every influence, every idea, has come from networks of hundreds, too many to mention. We are all interconnected in this way, so thank you to you all.

More personally, I will be forever grateful for the moral support and counsel from Anna Söderblom, Dr. Aleks Krotoski, Daniel Griffiths, Maya Hart, Lara Carmona, Kevin Slavin, Adam Greenfield, and especially Lucy Johnston. Incredible thanks to my mother and father, for foolishly leaving a modem in my room twenty-five years ago. And for Pico and Edwin, who took the brunt of the birth pains: that was horrible. Let's do it again.

00

CHAPTER ZERO

Most people who use the Internet don't even have a basic grasp of how it works. This isn't surprising: most people who turn on a light switch have no clue about the engineering, or indeed the scientific principles, that enable them to stay up long after sundown. And of course it goes without saying that the Internet represents a technological achievement of such complexity that it dwarfs almost any other in recent history. But it's not just the recreational users who don't get it. Arguably, if someone's only contact with the digital world is ordering from Amazon and reading their newspaper online, it doesn't matter much if they don't understand the technology that allows them to do so. They just do it. After all, an *X-Factor* viewer might not be able to explain to you how the broadcasting industry works, but that doesn't mean they can't enjoy bellowing at their TV screen and giving you a detailed appraisal of this year's runners-up. The thing is, though, that the Internet is absolutely *not* just fancy television, and actually it's my contention that it *does* matter. The Internet matters to all of us—that's why I'm writing this book.

The gulf of understanding between the technologically literate and the technologically illiterate has profound consequences. Most of those who are responsible for, say, writing and passing the legislation that governs our world have no more grasp of how the Internet works than our friend who checks the *Telegraph* online for the sports headlines. And since the Internet is changing every aspect of our lives at an accelerating pace, it really is a problem when our political leaders, or our pundits, artists or businessmen, don't get it. They make bad assumptions, pass unenforceable laws or invest money in unworkable solutions to the wrong "problems." And they are regarded as profoundly out of touch with the new reality by those who are busy crafting it.

This is not to sneer at those who do not yet get it—the digital environment often looks and feels dauntingly complex, and some of the philosophical, social, political and economic consequences of living in a networked world are hard to take in, especially for those of us who were not brought up with them. That's the aim of this book: guided exploration, in the hopes that it may help us all to feel a little more comfortable. You do not need to brace yourself for an avalanche of technical information. Even a little knowledge of the underlying structures that make up the Internet goes a very long way.

The origins of what we now call the Internet were designed for the U.S. military. The brief was to build a communications network between military bases that would withstand nuclear attack. Hundreds of physical locations needed to be in touch with each other. To connect them in a loop was out of the question, since an attack on any single base would take out the whole system. The obvious alternative (connecting every point to every other point) required impractically vast quantities of materials. The solution was ingenious. Every message that needed to be sent was to be broken down into "packets" of information, as if a letter were cut up and each piece posted in one of twenty different envelopes. The packets would be sent, not direct to their ultimate destination, but to the nearest available computer in the right general direction. Eventually, all the packets would arrive, by various different routes, at the end point, where the message would be sorted back into its correct order and opened. It's as if you decided to send a letter to your grandmother in Penzance by walking out your front door in Newcastle and asking passers-by if they were heading south and if so, would mind passing on a line or two of your message. By various hops, it would eventually all make it to Cornwall—at least in theory!

Thing is, the theory worked. The great advantage of this system was that because the packets of information weren't following any set route, if a bomb took out points D, E and N they could still make it from A to Z via myriad other paths. The process might slow down, but the information would still arrive. And if a couple of packets failed to turn up and the message could not be reassembled, the end computer simply messaged the originating machine and asked it to try again. The brief was fulfilled: anyone connected, even by a single link, could access all the other points on the network.

In that single fact is contained the spark of the revolution that has transformed our lives. So long as your point of connection is hooked up to your local network, which is in turn connected to the regional one, which is in turn connected to the national one, which is then hooked up to the transatlantic cable, you can video conference with your friends in San Francisco, Beijing and Buenos Aires in real time. At hundreds of thousands of stages along the way, individual computers called routers will make a decision about the best direction in which to send your packets of data.

The fact that this works is almost miraculous. It certainly feels that way to the end user—or at least it does when we remember to notice it. The Internet was planned by hundreds of engineers, scientists and visionaries working over the past fifty years. It took extraordinary amounts of work, idealism and determination to develop it to the point at which such everyday miracles as three-way video conferencing in three different time zones are possible.

And yet, for all its incredible sophistication, there are still weak points in the architecture of the Internet. Whole regions remain cut off from the global network. Bangladesh is connected to the rest of the world by just three fiber-optic cables. Before the World Cup in 2010 brought an urgent need to install more infrastructure, South Africa was connected via a single cable. Where this is the case, a region is very vulnerable to "unplugging" of the sort ordered by the Egyptian authorities during the popular uprising that overthrew President Mubarak.

But while these bottlenecks make restriction relatively simple, in engineering terms intervention is not a subtle procedure. A crucial thing that most technologically illiterate people fail to realize is that you cannot censor the Internet along national boundaries without applying draconian measures. The technical capacity to monitor—i.e. to decode and read—all the flow of information through those fiber-optic cables does not exist. You can turn it on and off like a tap, or set it to trickle mode, but you cannot filter its flow for individual elements. And since the Internet treats any kind of block (including attempts at censorship) as damage, it simply looks for an alternative route and sends its packets of information off in another direction. It is virtually impossible to prevent the flow of information, and this is true both on a technical level and also, as we will see, on a cultural one. A surprisingly high proportion of Western politicians have not grasped what their Chinese opposite numbers have learned through practical experiment: there is no such thing as light-touch censorship online. Nation states can only restrict what appears on the Internet within their jurisdiction if they are prepared to be draconian, and even then they will be only partially successful.

The censorship issue offers a good case study for talking about how people's lack of understanding translates to wrong-headed attempts to intervene in the digital world. Codes of voluntary agreement among Internet service providers appear to offer an alternative way to control content, one far more palatable to governments of a liberal disposition than turning off the tap, *à la* Egypt. For example, in the U.K. all providers have agreed to block child pornography sites, if they are brought to their attention. While no one in their right mind would disagree with the attempt to protect children, it's worth noting a couple of problems with this sort of voluntary agreement. Firstly, there's the fact that no service provider has the capacity to automatically block such a site in the first place if they don't already know about it, because to do so they would need to be able to accurately assess and filter all the information they pass along for its content. But information does not whizz around the Internet in easily viewable images or readable text. It travels in tiny packets of code. The sheer volumes of code make the process of deciphering all of it on some hypothetical master-reader completely impossible. To understand why this is so, it might help to remember that in order to be able to identify child porn you would need to be able to identify *every other single thing* that the code could conceivably be and eliminate it from your enquiries.

It is not the case that you cannot intervene in relation to online content, but you cannot do it by network censorship. The way to close down child porn sites is through police work: enforcement and entrapment. That fact is difficult for people who do not understand the architecture of the Internet to grasp. It has resulted in a lot of the sort of unworkable legislation that liberal regimes tie themselves up in when they are trying to enact their values from the unfortunate position of technological illiteracy.

The other problem with voluntary agreements is that they are not subject to any disclosure of information legislation, so we have no idea what other sorts of content is on the list of things the authorities do not want us to access. Child porn is at the extreme end of the spectrum, the nuclear option in the Internet censorship discussion—nobody's going to argue that we have a right to access it. But even if we personally are fine with trusting our rulers to decide which other things we can and cannot see, the Internet collectively is absolutely not fine with it. As we will see in later chapters, the Internet treats restricted access and closed data sources as a challenge, even a call to arms. This is another point where the divergence of opinion between the networked, techno-literate generations (i.e. those who've grown up using the Internet) and the hierarchical non-literate ones (those who haven't) is most acute. For anyone who lives their life on the Internet, it is considered rude when someone denies you access to something. The Internet was built and is still creating itself though a principle of collaboration. If you post something online that is restricted access, you are rejecting that collaborative instinct. You should not be surprised that your content is likely to be targeted by hacktivists—online activists who use their hacking skills to gain access to the very data that you've tried to lock away.

The key concept is that you cannot control the way people use the Internet without "doing a China" and that has been of only partial success, even for the Chinese. When the market acts as a censor, as it does when it restricts German Harry Potter fans' access, it is even less successful, because it is nowhere draconian enough to do the only thing that is in any way effective: turning off everyone's access to everything. Content-providing corporations, national governments and all the other old-guard elites that, in my view, are rapidly becoming extinct keep waiting for some new piece of legislation or technology to save them, to roll back the last twenty years. The fact is, that isn't going to happen. Once we all accept that, we can stop wasting our energies on panicking about the impact of the Internet and get on with reimagining our world. It's exciting and empowering to make that shift from hierarchical to networked thinking. If you are still feeling either panicky or confused, this book is full of suggestions on how you can learn to love the Internet.

01

MOORE'S LAW

When it comes down to it, the math is relentless.

Gordon Moore, a co-founder of Intel, the microprocessor firm, wrote a paper in 1965 describing a curious observation. Every year, for the seven years since the invention of the integrated circuit, the number of components used on a microchip had doubled while the price remained the same. This trend, he thought, might go on for another decade. By 1975, when he revised his prediction to every two years instead of every year, he had been proved correct.

Moore's Law has left its engineering background and entered the modern culture, and while there are various technical caveats and simplifications that might send a chip designer into nerd rage to read this, it is now more generally held to be the following: computers double in power every two years. Or, for the same power, halve in price.

While not a law as such, the observation has been true since 1958, and doesn't look like letting up any time soon. It is because of Moore's Law that the rest of this book is possible, and we need to understand its ramifications before we continue.

Firstly, Moore's Law makes planning very difficult. Imagine you have just been made Prime Minister. With good luck and a following wind, you might expect to be in Downing Street for eight solid years. But the technology policies you put forward in your first year in office will be based around technology that will be laughably obsolete by the time you leave. The mobile phone on which you took the congratulatory calls during election night will have been replaced by something sixteen times as powerful, and the most expensive phone on the market on day one will now be given away for free.

For buildings and city planning, Moore's Law presents an even greater challenge. If you consider a new building might have a usable lifespan of fifty years, the technology used within it as the wrecking ball is swung will be thirty million times as powerful as today. How, then, do you take this into account when you are designing the fabric of the building?

The long-term increase in computing power is breathtaking enough for a desktop computer, but consider the reverse effect. We are planning cities today that will one day hold technology more powerful than we've ever seen, smaller than we've ever seen, and so cheap as to be almost free. The idea of a supercomputer the size of a pack of playing cards in the shops for less than a few hundred pounds and made by the million might sound like over-optimistic science-fiction. Or at least it would have done five years ago, before Apple brought out the iPhone. Cities of the twenty-first century, as we'll discuss in a later chapter, may well be designed around the mobile phone in the way that cities of the twentieth century were designed around the car. Either way, it is our duty to ensure that the permanent infrastructure we lay down today takes into account the stuff we'll be putting into it tomorrow. That stuff is subject to Moore's Law.

So it also goes for our careers and our schooling. An

eleven-year-old will see a sixty-four-fold increase in computing power by the time she leaves secondary education. A career executive taking twenty years to reach upper management will be greeted by a technological landscape half a million times as powerful as the day he started. In a knowledge economy, this relentless thrust forward is the only constant we have, and as information technology touches ever more aspects of our lives, Moore's Law becomes contagious. What were once fields of human endeavour untouched by Moore's Law are now sucked into its upward spiral: the military, farming and culture of all forms are now shaped by the logarithmic graph.

The relentlessness of Moore's Law need not be tiring, however. It does provide an opportunity for slacking off. Let's say you have a job processing a large amount of data, enough so that if you start with today's technology it might take you six years to complete the task. If you were to sit on a beach for a couple of years, however, and then start with the (twice as powerful) technology available then, the whole task would take only three years. Even with the time spent working on your tan, you'd still finish five years from now, a whole year sooner than if you started today. Thinking of digitizing your entire CD collection? Wait a bit, and it'll be over sooner.

Moore's isn't the only law that describes the pace of technological change. Kryder's Law, named after the research engineer Mark Kryder, says that the amount of data you can fit onto a magnetic disk of a given size will double every year. In high-street terms, that means that every year a portable hard-disk drive of a given size will halve in price, or the same amount of cash will get you twice as much space. The effects of this are perhaps more striking than Moore's Law. There are easily guessable values for the amount of storage you would need to never have to throw away any photograph you took during your life, for example, or to store everything you ever read or watched or heard. You can work out quite simply the storage needs for every piece of music ever recorded, and Kryder's Law then simply points out the day when such an amount of storage will be affordable. As I write, one terabyte of storage, enough for around 200,000 songs, costs around fifty pounds. In ten years' time, the same price will give me storage enough for 100,000,000 tracks. A box the size of a hardback book containing, say, the entirety of Hollywood's twentieth-century output is technologically, if not socially, foreseeable without any effort. So too is a book-sized box containing every book ever written. In technological terms, we just have to wait.

Awareness of these laws also helps negate the temptation we might have to dismiss the first versions of technological advances—especially ones that look like they may threaten our livelihoods. The first iteration of a device, an idea or an online service is invariably rubbish, and it can be very comforting for the industry or group that is threatened by the advance to dismiss it entirely on that basis. The new rival is too slow, or too cumbersome, without enough memory, or with too low resolution a screen. This is dangerous. If you ever find yourself dismissing an idea because the first implementation isn't very good, then you must ask yourself if the implementation is being held back solely by the available technology. If that is the case, like our beach-bound researchers, you need only wait a while. Too many people dismissed the very idea of digital cameras because the first ones took bad pictures, mobile phones because the first ones were too big, or MP3 players because the first ones couldn't hold more than a few tracks. They were proven utterly wrong. A disruptive idea won't be stopped by today's lack of capability. If the idea is good at all, it will simply bide its time.

02

THE CLOUD

In years to come, when we think back, it will be the Cloud that will represent the beginnings of the twenty-first century. As the Internet becomes the dominant platform for our cultural, social, political and business lives, we find ourselves living in the Cloud. It is both a concept and a thing with a definite technical meaning, and understanding both of these is the beginning of our journey through the modern world.

Conceptually, the Cloud is the place you are when you're online. It's where all the information, the communication, the *stuff* is. When you send an email, it goes through the Cloud. When you're downloading a file, it's coming from the Cloud. When you have an instant messenger conversation you're sat in the Cloud, and when you become part of a community on a website, that community meets, talks, and prospers in the Cloud. It is the place where your mind is augmented by faster sources of information. The Cloud is, in the words of the author Cory Doctorow, an "outboard brain." It is, as we'll see later, just as real a place as the real world is, if intangible and fleeting and hard to grasp.

Technically speaking, the idea of the Cloud is perhaps less poetic. It is that a device with a fast enough connection to the Internet can treat other machines, out there somewhere, as extensions of itself. Your laptop might only contain a relatively tiny hard drive, for example, but it could be connected via the Internet to a machine with huge amounts of disk space. If your connection is fast enough, there's no need to think of this extra space as something separate and "out there" at all. You just think your laptop has suddenly grown a larger drive. In this way, the Cloud conceptually blurs the edges between all of our computing. With a fast enough connection, storage and processing power are effectively infinite.

The Web is a good example of this. You can access most of the Internet from your phone, even though it certainly doesn't contain its own copy of the Web right there in your hand. No one has that, and never will—Moore's Law notwithstanding. Instead you connect to the Cloud and access the sites you want from within those machines that make it up. Your phone is effectively as big as the entire Internet and as powerful as the most powerful machine it can connect to.

Another weird thing is that you have no idea where these machines are physically located; nor does it matter. Indeed, many websites have servers spread between different buildings or even different countries, and you will never know (unless you really care and have a very strong knowledge of network engineering) where they are. It just doesn't matter. On the one hand, this is quite sensible: a power failure, a natural disaster or a war could bring down one of your data-centers, but if you have more than one, your site and your business will remain up and working. On the other, it makes national laws very difficult to police on the Internet. A website owner under threat of legal closure in one country can physically move her servers to another, more casual, country—or, more likely, simply move the data from one machine to another through the Cloud—and Internet users won't even notice.

The way the Internet works means that when you come to draw it, as engineers often have to do, you can skip the middle sections entirely. The structure of the network between here and there is both irrelevant and constantly changing anyway. Drawing it would be pointless, and so it's common practice to abstract it away by drawing a big fluffy cloud shape. Hence the name.

The earliest Clouds were for storage, as I've described, but it soon became clear that you could use the machines making up the Cloud to do some real work. After all, a large room full of powerful servers can do a lot more computing a lot faster than your simple phone or laptop. If you are storing all of the data up in the Cloud anyway, then it's an obvious step to do the work there too. Web-based applications work in this way. When you ask a mapping site to plot a route for you, or set a filter on your webmail service, that work isn't being done by your browser. It's happening up in the Cloud.

This model of computing, with a "dumb terminal" at one end and a big machine at the other, is actually rather old. The original mainframes in the Sixties and Seventies worked in this way. Today though, the big machines are actually collections or clusters, hundreds of thousands strong, of simple machines connected together, and the "dumb terminal" is itself quite smart.

E-commerce companies pioneered a modern form of this. If you buy a book from Amazon, for example, your browser connects to a cluster of servers somewhere in the world, and the whole transaction, from browsing their catalogue to making the payment, happens in the Cloud. Over the first years of its business, Amazon built massive data-centers around the world to house the banks of machines that allowed them to provide their services. They built so many, in fact, that they were able to rent sections of them out to the public. Today, as well as books and other physical goods, Amazon has a core business providing other companies with access to a powerful Cloud. If you have a website that needs to store large amounts of data, or a program that needs to run on very fast machines, you can rent them by the gigabyte or by the hour from Amazon, or another supplier, and access them over the Internet. You'll just never see these machines, nor know where they are, and in fact the machine you rent will be actually be a simulation of your desired machine inside an even bigger one.

This is hugely significant. For twenty-first-century entrepreneurs, these Cloud services solve a very real problem: what do you do if your new business is a success? This isn't as silly an issue as it sounds. Computing infrastructure is expensive, and so no one wants to invest in big servers if their idea turns out to be a flop. Choose to only buy a small machine, however, and you run the risk of the site being brought down by its popularity. In the first years of the Web, a link from a very popular site could actually close you down as the rush of curious visitors overwhelmed your server. Slashdot, a discussion board for programmers and associated geeks, was so notorious for this that the phenomenon was named after it. Some of the sites that were slash dotted stayed down forever, especially if their owners were paying their server company by the amount of data transferred rather than by the month.

With the Cloud, however, you can launch your new business on the processing power equivalent of a tiny machine, and only pay for more processing power, or storage, or data transfer as and when you need it. If it's not popular, it won't cost you much to run, if anything. If it is a hit, you can ask your Cloud supplier to dial up the power to deal with it. Digital businesses can therefore be elastic—their costs can grow incrementally with their success. For the older incumbents in, say, the newspaper industry, a new business would require a massive outlay just to get started, buy the printing presses and so on. The Cloud is one of the reasons that purely digital businesses invariably destroy their older rivals, unencumbered as they are with the legacy of old machinery they can't be turned off when times are hard. This ability to adapt constitutes the digital world's asymmetrical advantage.

03

ASYMMETRY

Life in the West used to be simple and balanced. For every us, there was a them. The political right against the political left, the good versus the bad, company X competing with company Y. While each side may have entirely differing viewpoints, or moral codes, or commercial offerings, their very nature was in fact quite similar. Our enemies, until the fall of the Berlin Wall in 1989, may have been Communist, but they too had offices and bureaucracy, wages to pay, and an address where we could send them a birthday card. Your firm's commercial rival operated under the same strictures as you did, and although one of you might have a much better idea, and all of the success that comes from it, you were both playing by the same rules. This symmetry was a notable aspect to the Cold War world. It is the world in which the baby-boomer generation grew up.

After 1989, this changed. As the Warsaw Pact countries fell in on themselves, the West found itself without a big enemy.

Germany reunited in October 1990, and a few weeks later, on Christmas Day near Geneva, Tim Berners-Lee turned on the first web-server. Symmetry, as a factor of Western life, would be gone forever.

The Web, as we've seen in the intervening years, allowed a class of businesses to start that were fundamentally different to their older rivals. Where old businesses were restricted by geography, online businesses could serve the whole planet. Where old businesses needed huge investments in manufacturing simply to make the thing that their product was delivered on—printed paper, magnetic tape and compact discs in the case of the media industries—the Web needed nothing but a server and a connection to the Internet. While whole swathes of businesses once relied on an intermediary to simplify matters, today's web-driven world allows the consumer to connect directly to the thing they want to buy. This disintermediation, as it was called at the time, is why you have most likely not visited a travel agent in years.

The travel agents were hit by an asymmetric rival—the airlines themselves and the realization that it was easier to just book the flight yourself online. A similar fate fell on American newspapers. There, the majority of their income came from classified advertising: jobs, houses, things for sale, services offered and so on. Along comes the Internet, though, and the majority of this advertising switches to websites almost overnight. Craigslist, the most popular classified advert site in the U.S., is often credited with single-handedly (indeed, almost literally, as Craigslist has only around twenty employees) breaking the whole U.S. newspaper industry. If they did, it wasn't on purpose. They didn't set out to take on the newspapers. They set out to provide a nice service for people who had stuff to sell. It's not that they beat the newspaper industry. They didn't win. They weren't even playing the same game—and that fact is the part that creates confusion among many baby boomers. Not only is the world rapidly changing under them, but the traditional structures of business and society are being replaced by alternatives that have no interest in the old way of doing things. It seems motiveless and arbitrary, and to some might even feel malicious. Why, oh why did Jimmy Wales have to start Wikipedia to destroy the paper encyclopaedia industry? Well, he didn't. That's just what happened in his wake.

Asymmetry, then, comes from a rivalry between opponents who are completely different in both their style and substance, and also their intent. For sure, Internet companies are generally speaking blessed with lower overheads, faster development times, more nimble business strategies and less of a need for large initial investments than old-style firms. That they are much less restricted by national borders, by distance, by the time of day, or the difficulties in being discovered in the first place is all true, but they're also likely to be offering competing services for completely different reasons, many of which are not commercial in nature. If Craigslist stuck the knife into the back of the U.S. newspaper industry, it was enthusiastic amateurs, writing on their blogs about their own personal obsessions, who really twisted it—and they knew not what they were doing.

Add up all of the basic advantages of Internet-based businesses, and it is easy to see why the huge changes across business and culture over the past few years have arisen. If you are reading this in the print edition, and bought it by physically visiting a store, you have been on one side of an asymmetry. If you are reading it on an eBook, and bought it via a wireless connection, in the middle of the night, in the middle of the countryside, you are on the other. The asymmetry calls for a whole new class of commercial warfare, and it is this battle that we are watching today.

Asymmetry isn't new, of course. In the history of warfare, most battles were asymmetric. Agincourt, for example, had the old-school French knights losing badly against the newly gadgeted English archers. Over the twentieth century, and especially after World War II, asymmetric warfare became the norm, and just as the Internet was fixing itself as the dominant asymmetric platform of the modern age, the attacks of September 11th gave the best example yet: a tiny group of people with novel tactics inflicting terrible damage on the largest military society the world has ever seen. Asymmetry is the dominant theme for the modern age. For reasons we'll discuss in later chapters, Al Qaeda and the like demonstrate this in the political arena, while in the worlds of business and culture there are endless examples of Internet firms who have risen seemingly from nowhere to destroy whole categories of traditional industries.

The lesson to learn is that apparent size, strength or position are not meaningful in the modern digital age. The little guy can always beat you. Let's look at why.

04

THE SOCIAL GRAPH

With perhaps a tenth of the planet using Facebook every day, less than a decade after its invention, social networking—and the idea of the social graph—is perhaps the most influential and culturally significant thing to have happened to the Internet. We'll be discussing its effects in many chapters in this book; but before we can do that, we have to understand the basic idea, and just what a social network is for.

Despite the official history, and the rather good film, the roots of Facebook aren't in Harvard, or even in Silicon Valley. They are instead to be found in the writings of the Hungarian author Frigyes Karinthy. In 1929, Karinthy wrote a short story, "Láncszemek," in which he propounded the idea that everyone on the planet was connected to everyone else by no more than six degrees of separation. That phrase then became the title of a play by John Guare, propounding the same theory: that we are all connected by short chains of acquaintance. I'm connected to you, dear reader, because I know someone who knows someone who knows someone who knows you.

While the idea was subject to experiments, with varying success, its veracity doesn't really matter. Simply the idea that we might be able to contact—be friends with—anyone on the planet through a series of introductions or personal connections is very pleasing. Our interconnectedness is important to us, especially when we consider our specialist field or the industry in which we work.

There are many ways to measure this sort of interconnectedness. A popular early website, The Six Degrees of Kevin Bacon, lead to actors having a Bacon Number, which was calculated by counting the number of links between film roles that join them to the actor Kevin Bacon. Likewise, mathematicians can work out their Erdös Number, based on authorship of academic papers that eventually link them to the Hungarian mathematician Paul Erdös. In both of these measurements the lower the score the better. For the truly connected there is also the Erdös-Bacon number, given to those who have both kinds of score. For example, Natalie Portman, the actress, having starred in a film with Kevin Bacon, scoring 1, and coauthored a mathematical paper that gave her an Erdös score of 5, therefore has an Erdös-Bacon score of 6.

Online social networks, then, started as business networking tools. One of the first was even called Six Degrees. The idea was that if you loaded your address book on to the system, and everyone else did as well, then it would expose all of the first-, second- and third-degree connections that you and your colleagues all have. So your friend, looking to talk to someone important, might ask you for an introduction to someone, who might then introduce them to someone else, and so on up the chain to their desired mark. Social-networking applications were once exactly the same as traditional cocktail-party business networking.

That usage still remains, but today the social networks, and Facebook in particular, have become platforms on which many other services are built. Now the idea is that the social network not only allows us to connect to specific people through our friends—whether that's the President of the United States or Kevin Bacon—but that it can also tell other people, advertisers, something about our likes and interests. How can this be? Well, when we declare who we are friends with, and we and our friends declare what culture we like, the social network can go a long way towards identifying our individual wants and desires. The interests of our friends are considered to be very good indicators of our own interests, even if we don't explicitly state what our own "likes" are. This is very compelling for advertisers, whose main problem has always been finding a way to advertise solely to the people who are receptive to their specific message.

This is why Facebook is designed to be so compelling. The services that Facebook provides are very useful to their audience—messaging, online chat, photo sharing and so on—and fun and entertaining, and there is also a deep-seated need for many people, young people especially, to define themselves to the world by listing their interests and registering the things they like. Every additional data point you volunteer goes to help make the advertising that funds the site ever more effective. It is instructive here to remember who is the customer and what is being sold. In the case of Facebook, the customer is the advertiser, and what is being sold is you, the audience.

One of the common complaints about social networks, especially from concerned parents and teachers, is over the number of Friends that people have. It's not uncommon to have hundreds, even thousands, of Friends on Facebook, and of course there's no practical way, the complaint goes, that these are actually your friends. They're not the people you will ring in a tizzy at 3 a.m., nor the sort of friends who would, say, help you move house. They're not real friends at all. That is both true, and also missing the point. The Facebook capital-f Friend is a different concept to the lower-case-f friend. While there is an overlap, the group of capital-f Friends is also made up of people you met once, people you used to know, people who are friends of friends, or are in the same circles, and so on. They're loose ties of like-minded people, and the reasons for friending someone are many. Because some social networks share contact details—LinkedIn being the most current example—a friending there of even the most fleeting business contact allows both parties to keep their address books up-to-date automatically. Other social networks are simply fun to be on as a platform to show off, or to be witty, or incisive: friending happens liberally in order to grow each others' audiences. Twitter is currently the best example of this sort of thing. These groups of online contacts are not so much a replica of our real-world social networks, as the original idea was, but rather a very loose personal community of like-minded peers.

Real-world social networks, made of people we know and meet and interact with in the flesh, are different. The British anthropologist Robin Dunbar has postulated that the maximum number of steady interpersonal relationships that the human brain can keep track of is around 150, known as the Dunbar Number. He reached this number by observing that the social circles of primates varied in proportion to the size of each species' brain: the bigger the brain, the more friends the ape had. Extrapolating to human-sized brains, he came up with the Dunbar Number. Further studies has him saying that of that 150, perhaps only forty relationships can be meaningful in any way.

Stripped to its heart, then, a social network is an application where we can declare our interests and make public a special form of social relationship. For the users, the social network allows us to talk with our Friends, and engage in all of the conversation and sharing and mutual play that makes life worth living. For the businesses themselves, the users leave behind huge amounts of interconnected data, and this data can be put to very good use. To truly do that, however, they need a technology that can understand it. Which brings us to our next idea: the Semantic Web.

05

THE SEMANTIC WEB

Search engines are actually quite stupid. While it is ridiculously impressive to receive results from Google within milliseconds for any term you wish to search for, those results are limited to matches of the exact phrase you've typed in. Search engines can't infer things that aren't explicitly stated in one place. For example, although my name is on the Internet on pages associated with books I've written about the technology known as RSS, and also on pages noting that I have an Italian Greyhound as a pet, and I've also favorited a lot of videos of musical theater on YouTube, putting "person who wrote about RSS who has an Italian Greyhound and who likes musical theater" into Google won't bring back my name, at least not until this paragraph is published online somewhere. A human being, however, can make those inferences very easily. We do it all the time in conversation, especially in gossip: "You know, the girl at the party with the silly hat. Remember her? Well, she married that other guy. The one with the big car." And so on.

Indeed, most conversations have long chains of these inferences. Booksellers, for example, might be asked for a suggestion for a book "a bit like" something, but "more" something else, or a cinephile might want to see a new film directed by someone who had previously once worked with a famous writer. You can't Google for that: unless there's a page that explicitly declares a fact, modern search engines can't know it.

There is a solution to this, however. The Semantic Web is the idea that every written fact on the Internet could also contain a machine-readable version that search engines and other programs could understand and make inferences from. This machine-readable description of the page is written in a language known as RDF, or Resource Description Framework.

RDF can be written in various ways, but the theory is always the same: the assertions it makes are in the form of "triples": subject, then predicate, then object. So a page about my Italian Greyhound Pico might make a series of statements in RDF like so: Pico hasSpecies Italian Greyhound. Pico hasColour brown. Pico hasOwner Ben Hammersley.

Then another page, on a different website, might have statements such as Ben Hammersley hasDateofBirth 3 April. Ben Hammersley hasNationality British.

If a Semantic Web search engine was to index both those pages, then someone could ask it "Italian Greyhounds whose owner was born on 3 April," and the answer would be "Pico." This isn't directly stated anywhere, but simply inferred from the facts that are stated in a machine-readable way around the Web.

Supporters of the Semantic Web point out that this would be intensely useful, and much closer to the sort of questions we ask each other all the time. A question like, "Who makes the shoes that the lead actress wore in the film on the TV last night?" is impossible for a standard search engine to answer. A Semantic Web search engine would have no problem, provided that the data was available online.

To put the data online, the Semantic Web requires that pages are "marked up" with data, and that this data is itself both true and correctly described. All three parts of a semantic triple have to be solidly defined—just what does "hasSpecies" mean? Or "Italian Greyhound" for that matter? As these things are meant to be read by machines, one can't rely on context to work out the meaning. So the word "title" in English might refer to the name of a document, or an honorific given to a person, and when we speak to each other the meaning is clear from the context. Computers can't do that; so, in order to fix this, the Semantic Web is made of different vocabularies with commonly agreed-upon meanings. You end up with triples like this: ThisBook HasTitle(books) *Winnie the Pooh* and ThisMan HasTitle(people) Emperor.

There are many of these vocabularies, each defining different types of thing. There are vocabularies to describe geographical information, to describe video clips, even one—called "Friend of a Friend," or FOAF—to describe personal relationships. Ben isFriendsWith Dan, for example. A semantic search engine that understands these vocabularies can make inferences across them, linking pages that use FOAF to other pages that use a vocabulary to describe music, to another with personal details, and you can ask computers for very complex things: "List all the music listened to in the past three days by people who are friends with Ben, by artists who were born in October."

This data is already all out there, and is being slowly marked up in a semantic way, so we might be seeing forms of the Semantic Web, at least in specialist areas, very soon. However, there are problems. One major criticism of the Semantic Web is that it fails to encode doubt or mistrust in the data. Asserting a false fact in conversation—lying—is dangerous enough, but asserting a false fact into the Semantic Web makes for longer term and wider ranging problems. Whole systems can be built resting on the empirical truth of a single data source, and without a way of saying this source might be dodgy, the whole exercise might be terminally flawed. For this reason, to date the majority of Semantic Web projects have been using closed data sets—they're working on systems where everyone trusts the source of the information, and where the definitions of the vocabularies are agreed upon. This is, notably, the very opposite of the ordinary Web we use every day.

The Semantic Web does bring us to our next concept very nicely, however. You need to be able to point to what you're describing. Everything has to have a name, both to be able to refer to it, and to be able to differentiate it from other similar things. In a semantic triple, each different object is referred to by something called an URI, a uniform resource identifier. These look very much the same as an URL for a webpage. Each URI, however, is the identifier for the thing itself. I personally have a URI of http://benhammersley.com, so in the case of my dog, we can say Pico hasOwner http://benhammersley.com. In fact, I have many URIs—my webpage is one, my email address another, my phone number, as we'll discuss later in this book, is a third.

So far so technical, and you might be forgiven for glazing over at this point. But you'd be wrong. We use the concept of the URI, for people at least, increasingly more often. In Twitter, for example, my URI is @benhammersley. That's both how you would refer to me in general, but also how you refer to the me that is the me on Twitter. It's separate because the me on Twitter is different to the me on LinkedIn, and both of those are different to the me in newspapers, or in this book. We're starting to talk to and about people online in a way that acknowledges and separates the multitudes contained within each individual, and all of this comes from the URI we choose to use to point to someone. We have gained a whole new set of names. In the next chapter, we'll discuss how this works and what it means.

06

TRUE NAMES

We are referred to by many names. Hi, I'm Ben; but I'm also ben@benhammersley.com, @benhammersley, http://benhammersley.com and https://www.facebook.com/ben.hammersley. I have a name on Skype, a number on the phone network, and a series of names on group weblogs and message boards that I'm not going to disclose here. Although it was only the first that I was given by my parents, the others are still the names I'm known by in different areas of my life. They represent me, or rather a part of me. They're the signifier for the bit of me that is exposed when I'm in the situation where I use that name. This sort of pseudonym isn't unusual or new: *noms des plume* and *noms des guerre* have been around forever. But they've previously been restricted to the important or slightly unusual. Today, however, almost everyone online has a pseudonym representing a specific aspect of themselves, from differing home and work emails, to the macho executive also known on specialist forums as Fluffybunnykins1973.

This ability to have new, unique, private names online is one of the great things about the Internet. The freedom to adopt a new identity to discuss a sensitive subject, or to try a new persona on for size without it being connected to the "real" you, is now an important part of growing up.

Anonymous or not, online identities allow people to relate to you in specific ways. This can be tied in with the communications mechanism as well: conversations I have with people over email to my professional address are in a different tone entirely to those I have over instant messaging using my personal IM identity, even if it's with the very same people. We have different names on those services, and so are different personalities.

One tricky aspect that the future will have to learn to deal with is the etiquette of what happens when these identities start to merge. This is a risk, especially with something like Facebook, which forbids the use of fake names. Young graduates entering the job market this year are likely to have been on Facebook for their entire university life. Their identity as a drunken student, so thoroughly documented online, might clash with their new identity as a hopeful civil servant or merchant banker.

In 2010, Google's then CEO, Eric Schmidt, said in an interview that in the future young people will be allowed to legally change their name, simply to disassociate themselves from the collected evidence of their past—evidence that companies like his own are dedicated to revealing. There were some reports that saying this sort of thing resulted in Schmidt moving out of the CEO position; in this case his reasoning was based on a genuine problem: that society has yet to catch up with the trail of stuff we leave under our true names.

The risk of leaving bad things under our true names can be used to an advantage by others. Anyone looking at online message boards, especially those run by newspapers and covering the news, will be aware of the ugly tone that they invariably sink to. The quality of online discourse can be much worse than anything found even in the roughest of pubs. This is due to the Online Disinhibition Effect, a phenomenon that combines the safety of anonymity with the lack of social cues given by a webpage to allow ordinarily sensible people to say things they never would in the physical world. Because you can't see the hurt in the other person's eyes, or them clenching their fists, online discussions simply tend to disintegrate. There is a rule on the Internet known as Godwin's Law, named after the Internet activist and lawyer Mike Godwin. It states that "as an online discussion grows longer, the probability of a comparison involving Nazis or Hitler approaches 1." Message boards that allow anonymity become Godwinned very quickly indeed, which is bad, and so many message boards are now insisting upon tying your identity there with an identity elsewhere. The idea being that you won't be as anonymous, and so the tone of the debate will improve.

Anonymity online isn't always a bad thing. There are many situations where you would not want your birth name associated with you online: everything from enquiring about curious medical conditions, to exploring your sexuality, to living under a repressive regime would obviously be better done without a connection to your real self. But it is here that we find ourselves in need of new words to describe these identities. The Internet, and especially social networks, has created opportunities for us to become different people, complete with a different name. The twenty-first century will be the one where we have to learn to manage our different identities very carefully. But to say that these identities are different from our real one isn't entirely true. These are our real selves, or at least part of our whole, and now it is perfectly common for people to have long-standing, deep and meaningful relationships both online and off with people who only know them by their online name and identity. As the social-networking sites try to mine our activities for useful data, and as everything we do under any particular name is made ever more public, pseudonymity will seem quite sensible.

07

ONLINE DISINHIBITION EFFECT

We've talked briefly about the fact that it's easier to voice extreme and offensive views from the safety of your laptop than it would be in person. This is such a common phenomenon that we rarely take the time to think seriously about its ramifications: hence this chapter.

There can be very few sites dedicated to news and current affairs that don't have a comments section—and thus that unhappy ability for the reader to place their own views and opinions underneath the story in question. This is perhaps a sign of a greater societal shift. In the twenty-first century, rather than passively consuming culture, everybody expects be able to take part. Whereas traditional mediums of public debate—such as writing a "letter to the editor," for instance—are bound by fairly strict rules of etiquette, online discussions tend to become rapidly polarized and shouty. This seems to happen regardless of the topic of conversation—politics, sports or technology, say—hence the rule of thumb known as Godwin's Law, whereby the longer an online conversation lasts, the more likely it is that an opponent will be compared to Hitler and their "crimes" put on par with the Holocaust.

Nazi comparisons aside, we have grown accustomed to the fact that the tenor of conversation on the Internet is completely different from the tenor of conversation that takes place in other public spaces. You rarely get complete strangers hurling abuse at each other in a city park, for example. That said, if we stop and look at any form of political conversation we can see the roots of this online vitriol—especially in American politics, where the two sides are so polarized. Here the online component is particularly fascinating since Democrats and Republicans are so mutually opposed to each other that they don't even visit the same sites, or talk on the same message boards, and yet still manage to be horribly offensive to each other remotely.

Intensive research has gone into investigating the differences between conversations online and conversations in the real world. This has lead to the widely held belief that online conversations are subject to something known as the Online Disinhibition Effect. The theory goes like this: when you're sat at the dinner table, or in an office, or even a political chamber, and arguing with somebody, you're getting an awful lot of information about the effect that your argument is having on them. You can see the look in their eyes, the way that their face is changing and their body language. You can see that if you say something hurtful they look hurt. Say the same thing online, however, and you just don't get this sort of feedback. In fact, you have very little information to go on at all. Sarcasm can have a very different effect online than in person; as too can robust debate. What you might think of as a strong but fair political argument might well come across as absolutely aggravating to someone who disagrees with you because there is no mitigation of this emotion given by seeing each other as three-dimensional human beings. Thus arguments tend to escalate because personal agency is rendered invisible.

One comforting factor of online discussions is that you get to control your own degree of anonymity. Sometimes a news story will prompt a little flurry of discussion with your friends or colleagues or follow politicians, over Facebook or Twitter, say. In these situations you tend to know the people you're talking to, both online and off. At other points—for example, in a special interest forum or in the comments section of a media outlet—your words and deeds are identified solely by a made-up name. Essentially, there is nothing connecting your words to you. This frees you from social responsibility and you can say anything you like because it will never get back to your loved ones, to your colleagues or to your boss. On the one hand, such anonymity frees you from the tyranny of being held accountable for a few badly chosen words; on the other hand, there's little to stop you saying something really quite offensive. This is why over the past few years many national newspapers and online services have started pushing people to use their own names when creating user accounts, or at least supply details that are in some way connected to their real-life identities; for example, their credit card details. Even this very tenuous connection seems to be enough to give people pause before they say something truly offensive.

One of the most successful online message boards and communities is called MetaFilter. As part of the sign-up process, you're required to pay five dollars before you can have your user account. Even though your real-life personal identity has never been disclosed to the rest of the community, the combination of the five dollar entrance fee and the fact that the system administrator knows who you really are means that the community is by and large very well behaved compared with other sites. As we'll see in the Anonymous chapter, the difference in the conversation styles between the sort of message board where you have to say who you are and the message board where you're forbidden to say who you are is very marked. On the other end of the spectrum, sites like Facebook and Google are all about recreating and expanding real-life identity online. However, as we'll learn, this is nothing to do with promoting and improving the quality of online discourse. Rather it is to allow them to sell their product—us—to the advertisers.

But that's not to say there are aren't moderating influences in place, as we'll see in the next chapter.

08

COMMUNITY MANAGEMENT

It is often said that the Internet is, for many people, simultaneously a source of anxiety and part of the furniture. The extent to which we have grown blasé about using at least the more widely accessible applications on the Internet is extraordinary. We all email. Most of us read or watch content online; many of us use social-networking sites. Very few of us did so fifteen years ago. Facebook, of course, hadn't even been invented fifteen years ago. The dangers of not really understanding a technology that you use every day to run your life are obvious, and I hope that this book will contribute to the cause of dispelling the fog and zapping the fear. The temptation to think of the Internet, or perhaps more accurately the World Wide Web, as a Wild West frontier, is understandable but also overblown. One interesting nexus of the completely new but the not at all anarchic forces that do in fact characterise much online activity is the community moderator, or manager.

A community manager on any site where people congregate

and respond to one another in open forums is a force for good in the digital world, a bulwark of civilization. To understand why, we have to remember two things: firstly, there is an exciting new class of business that operates solely on the Internet that is enabling people to write the content that they themselves consume. And secondly, because people are inherently not very good (yet) at coping with the heady freedoms offered by Internet-based interaction, all too often they fall victim to the Online Disinhibition Effect and start behaving badly.

Unless you have spent time in an unmoderated or badly moderated forum, it is hard to imagine how unpleasant such a place is to be. Imagine a *Guardian* strand discussing George Osborne's school days, multiplied by the *Daily Mail* community going to town on climate-change scientists who choose to be single mothers. Narrow-minded, bigoted, elitist, racist, misanthropic and misogynist lunatics abound, and in cyberspace you can shout at them as loud as you like. Except if you do, of course, the levels of hostility swilling around are eventually likely to diminish the experience for everyone.

The role of the community manager is to enable a potentially free-wheeling chaos to function in a way that allows meaningful and rewarding interactions among users, so that they will spend more time on the forum. A community moderator can be anonymous, or tagged, authoritarian and prone to disciplining offenders, or calming and cajoling. They manage, steer, nurture, or censor the discussions online (depending on your point of view). They keep the peace and calm people down. They respond positively to particularly interesting posts, and scour the comments for content that might be of commercial value to the business that employs them.

A good community moderator—one who can steer a forum delicately, without clumsy intrusions, so that the benefits to its users are maximized—is like a great maître d' or the ultimate university librarian. These are not the people creating the food or writing the books, but without their abilities to act as an interface between producers and consumers—or in the digital businesses, between individual co-producers / co-consumers—the whole relationship would fall apart and the enterprise would run aground.

These people are highly revered by the digital communities they serve, and by their employers. A good community manager is like gold dust. Because if your business, or project or whatever it is, depends on people returning to your site on a regular basis and uploading their content (Wikipedia, Twitter, Pinterest) you need them to feel safe and entertained and listened to and valued. That's a moderator's job.

There are numerous other jobs that didn't exist fifteen years ago: many of the techie coding and design companies that now employ tens of thousands of people worldwide are new businesses that postdate the dot.com boom. But these engineering and design jobs are not of a completely new class of work. There have always been designers and engineers working in new technologies. They are the people who make new technologies happen. But never before have there been people employed to take care of the co-creators of a business, because never before have there been flourishing businesses where there is no product to be sold, where the business doesn't make anything beyond a space for people to do their own thing. Facebook is like a self-replenishing oil well, a source of power that gets stronger the more it is used but is not responsible for making the thing that makes it strong.

People who think that the Internet tends towards anarchy are right in some ways: on an architectural level of network design, it certainly does. Network neutrality ensures it. But as the chapters on copyright reform, hacktivism and open-source education will show, there is a fierce ethical code that governs the Internet too, a home-grown one derived from an ethic of collaboration and openness.

There is also a far more old-world virtue that is highly valued all over the Internet: courtesy, respect for other users, their views and their time. The people who administer Wikipedia, who chase down spam bots, who comb a newspaper's comments for trolls and ranters and haters of all kinds, these people are providing a service that we should all appreciate. There's no doubt about it: a community manager's job is a noble one . . . although perhaps lacking the cut and dash of the fighter pilots we'll look to in the next chapter.

09

THE BOYD LOOP

One of the most influential military strategists of the twentieth century was U.S. Air Force Colonel John Boyd. His most famous theory came from looking at how fighter pilots could win dogfights. A dogfight between two jetfighters is dependent on two things: the skills of the pilots and the capabilities of the planes themselves. For Boyd, tasked with both teaching the pilots and advising on the design of new fighter jets, the problem needed to be broken down into phases. Just what was going on in the heat of battle?

Boyd's answer was that the pilot is continually going through a loop of four different phases: observe, orient, decide and act. First you observe what the enemy is doing. Then you orient that action with what you know about the enemy and try to work out what they're doing, based on your prior knowledge of their psychology. You decide what to do, and then act, bringing us back to the beginning of the loop, observing again the enemy's reactions to you. This loop is also known as the

OODA loop, and there are actually two in action: yours and your enemy's.

This might seem like a simple description of what is going on, rather than a tool to use. But Boyd pointed out that there were two weak points that, all things being equal, might be taken advantage of. The first is speed. If you take less time to observe, orient and decide than your opponent does, then your actions will start to short-circuit his own loop. He'll be forced to stop his actions simply to have the ability to observe yours. He'll have to skip forward from part four of his loop to part one again just to keep out of your way. The faster you go, the more likely he is to make a fatal mistake.

The second strategy is being weird. Your opponent has to spend time observing and orienting himself, so why make that easy? An unpredictable opponent is perhaps worse than a fast one, because you aren't able to make guesses about what he's about to do based on what you would do if you were him.

The tactics inspired by these insights were very successful for Boyd and the rest of the U.S. military. They inspired the commissioning of fleets of lighter, more manoeuvrable fighter jets, and a change in the training of pilots, and extended out to land warfare—Boyd used his observations to guide the strategy for the first Gulf War.

The Boyd Loop isn't just for the military, of course. The insight is just as useful in sport, or business, or diplomacy, and certainly in the faster moving digital industries. In this way, a theory that was first developed for fighter pilots has found its way into Silicon Valley, and design principles of the dot.com start-up. The average technology company operates on a very fast Boyd Loop. Their processes are specifically designed to be permanently in the state of a Boyd Loop, with the option to act by changing their product very quickly. Web-based applications are especially fast around their Boyd loops: they can observe, orient, decide and act in a matter of hours: releasing tiny improvements almost immediately, and watching to see what these changes do to their customers' experience. This makes the applications feel as if they are permanently unfinished, but always improving. Originally, a piece of software that wasn't quite ready, but had been released to a small group of people for their feedback, was said to be "in beta." Today most web applications are in a Boyd-Loop-inspired permanent beta.

Of course, all things are not equal. Planes, people, organizations, whatever you are applying the Boyd Loop to, do not all have the same capabilities to observe, orient, decide and act. Some are blinded by their design, or their hubris, or simply through not looking.

Orientation is very easily derailed, through a lack of understanding, a mismatched world view, or even through simple denial. The feeling that this simply can't be happening to me right now is paralyzing, even for just a second, and that might make all the difference.

The solution to these problems lies in being better at observing what is happening, being equipped with a world view that can interpret it correctly, and in being capable of processing anything that comes along, whether it's simply weird, or emotionally provocative. Indeed, it's better to actually expect weirdness and provocation, rather than be surprised by it and stunned out of your Boyd Loop.

And for weirdness and provocation, where best to look next but China, and the *shanzhai* movement?

10

SHANZHAI

Fake goods have been around for centuries. There has always been an desire for cheap imitations of otherwise unobtainable products. In the past decade, as China has further industrialized, the market for fake, or *shanzhai*, goods has grown to be significant.

Shanzhai, whose translation in Chinese means "mountain stronghold" and gives a sort of Robin Hood image of taking from the rich, is more than just making a dodgy knock-off. New manufacturing techniques, a growing consumer culture, and a disregard for local regulations and international intellectual property laws have meant that Chinese *shanzhai* manufacturers have become experts in not only copying fashionable Western goods, but also improving on them. They are thus placing a fat finger on the fast-forward button of product evolution.

The most common *shanzhai* products are mobile phones, but there are *shanzhai* versions of sportswear, tools, even architecture. In July 2011, perhaps the most beautiful of all *shanzhai* exploits was discovered by an American blogger, the pseudonymous "BirdAbroad," in the remote and relatively small Chinese city of Kunming. There, in a pedestrian shopping district, she found not just fake Apple products, but an entire Apple Store. Apple's genuine outlets are highly designed environments, with very controlled branding—they have the same tables and displays and architectural fittings the world over. The Kunming store had all of this, plus the correct employee uniforms, staff name badges and promotional posters. The staff themselves even believed that they were truly working for Apple. They weren't, and neither were the staff in the two other Apple Stores found in the same city. In an interview with the *Toronto Star* newspaper, the manager of the store claimed that while he admitted the store was fake, the goods they sold were genuine Apple products, though he declined to say where they were from. The *shanzhai* in question here, then, wasn't the iPad or the iPhone, but the shopping experience itself.

Experiential *shanzhai* isn't restricted to retail outlets. You can think bigger than that. Take the case of the Austrian village of Hallstatt. Home to fewer than a thousand people, and nestled on the shores of the Hallstätter See, the village is so picturesque that UNESCO made it a World Heritage Site in 1997. In 2011, news broke that among the 800,000 tourists the village sees every year there had been a team of Chinese architects with a plan, and now the Guangdong region is reportedly to soon have itself a exact replica of Hallstatt, albeit with the subtropical weather north of Hong Kong.

But even this isn't unique in China. Around Shanghai there is Thames Town, a village built to resemble an English market town, complete with a facsimile of a chip shop from Lyme Regis, cobbled streets, red telephone boxes, even the double yellow lines forbidding parking. It's one of a set of replica towns planned, with others in Swedish, Italian, Spanish, American and German styles. Whether anyone will actually want to live there is another question entirely.

While a *shanzhai* village might be pushing it, *shanzhai* products are no longer simple cheap knock-offs. *Shanzhai* manufacturers are subject to the same market forces as everyone else, and very soon they started to modify their copies to better suit the needs and culture of their local customers. Typical *shanzhai* mobile phones, for example, have slots for two SIM cards, and hence can receive calls on two numbers, where the original only has one. In late 2011, Nokia was forced to release its own version of the *shanzhai* versions of its own older phones—with the two SIM slots—simply to stay in the market in China. *Shanzhai* iPhones have FM radios and memory card slots, and come in different colors to the Apple-made genuine products, all because those are the features the Chinese *shanzhai* consumer is asking for.

Shanzhai labels are becoming important brands themselves, and starting to expand abroad. The Adivon brand, which is remarkably similar to Germany's Adidas brand, even bought advertising space next to the court during an NBA basketball game in early 2011.

Shanzhai manufacturers, especially in the mobile phone industry, are rapidly taking over markets in the developing world. Because *shanzhai* firms have no overheads for vast marketing campaigns, or rent to pay on huge HQs in Western capitals, their handsets can be much cheaper than the official versions. They can also be made in very small batches, which means that any innovation can be incorporated and on the street in days, if not hours. In societies where the mobile phone is a primary signifier of social status, the ability to upgrade your handset to something even more impressive every few weeks is very appealing, and the *shanzhai* manufacturers can tailor their models to the local fashions. The latest Nukia, Blickberry, or other cunningly misspelled knock-off will be more likely to do what you want, in the way you want it, than anything that has gone through twelve months of corporate business development in Helsinki or Waterloo, Ontario. Chinese *shanzhai* phones are very popular all the way across the developing world, even as far away as the Middle East and North Africa. The phone-facilitated revolutions in Egypt and the rest of the region were in no small part enabled by *shanzhai* handsets, designed and made by a country with its own totalitarian regime.

The iterative development process shown by *shanzhai* manufacturers allows them to work faster, and release counterfeits earlier than their Western rivals can get the genuine product on the shelves. A simple website photograph of an official product will become a *shanzhai* version in a matter of days, and improved models will be out shortly afterwards, all while the mainstream manufacturer is still clearing up from their press launch. This speed of innovation, and speed to market, is a major challenge for mainstream manufacturers in the developing world, Western or not. In an odd way, the knock-offs might not be as good, but they are better. Either way, they are certainly faster to appear.

11

CHARTER CITIES

People naturally want to go where they can most prosper. But those places are often reluctant to let them in. This we know. What is talked about less often in the developed world is that the places they come from often aren't very happy that they're leaving. It's a bad thing for a country when the most talented, the cleverest, the most ambitious of its citizens want to leave. For this sort of economic migrant, the root cause of the desire to go somewhere else is that the new place runs under a different set of rules: the collection of liberal trade laws, anti-corruption measures, freedom of speech and business and so on that will allow them to prosper. This isn't simply the large macro policies—liberal democracy versus theocratic regime, for example—but also things like employment regulations or a liberalized market for electricity or bandwidth or the sale of goods. Families want to live under a regime of public health care and education; businessmen want to live under rules governing fair play and contracts and property; everyone wants to live where there is no crime. Sadly, the leaders of the origin country often find themselves in difficulty. They might want to change their country's rules to match the destination country's, but local politics might make it impossible to do so. Meanwhile, people keep leaving.

The answer to this, according to the economist Paul Romer, is the concept of Charter Cities. The idea is simple. First, find some uninhabited land within the origin country, big enough to hold a city. Second, decree that a city built there will operate according to a new set of laws: the charter. Third, build the city and invite people to live and work there.

At first this idea sounds silly. For an Old European like myself, the idea of building a city of millions from scratch sounds like madness. It's not: new cities are being built across the globe all the time. Secondly, the idea of a city running under different laws and systems to its surrounding country might sound weird, but it's not really. In fact, it has happened a lot. Take China, for example: Hong Kong is part of China, yet runs under different rules. So is the neighboring city of Shenzhen, and nearby Macau. There are more than 3,000 Free Trade Zones across the planet, where countries waive import and export tariffs or provide tax breaks for companies coming there to set up.

But Romer's ideas are even more daring than Free Trade Zones. Those are usually nothing more than over-built massive industrial estates, still operating within the laws of the host nation. Charter Cities are complete cities, with all the infrastructure, culture and non-business buildings that implies. Most surprisingly, he suggests that a country might invite another country to run the Charter City—or at least supply the new laws, and train the new authorities within. Instead of, to give a fictional example, Guatemala watching her best people leave for Canada, Guatemala might invite Canada to help set up a Charter City within its own boundaries, implementing Canadian laws and regulations within the city's borders, training the police and civil service and monitoring them to make sure they remain obedient to their charter. Guatemala doesn't lose its best people and thus gets wealthier; Canada gains a whole new base with which to trade and interconnect; and the world is richer thereby.

The whole concept creates a series of new marketplaces. Within countries with Charter Cities, people would have a choice of rules under which to live. The cities themselves would compete with the rest of the nation. There needn't be only one: a country with a lot of space might build a Charter City based on U.S. law over here, and one based on German law over there. This would allow the host country to experiment with new systems and laws in a way that was much more acceptable than simply transforming their entire nation in one go. There's a second market too, in the rules themselves. Developed countries could compete with each other for nations to ask them to provide the rules for their charters. It's really the ultimate form of branding: we liked the United Kingdom so much we copied their entire political and social system. This in turn might have an effect on the supplier nations: they would need to streamline and codify their system in a way that would allow it to be sold to another country. The unwritten United Kingdom constitution would be hard to promote. Written down, it might well be a huge success.

Critics are not short of things to say about the idea of a Charter City. They're neocolonial, they say, and pay no attention to existing power structures within the country. It's all very well having a shiny new city, with lovely roads and buildings, if the people who move there bring their old corrupt ways with them. The need for the independent oversight of the rule-giving country would possibly cause friction. Furthermore, you'd have to be careful that a Charter City didn't leach money and talent away from the rest of the country in a way that was actually harmful to the nation: making a Charter City tax-free, for example, would be foolish, as all the good businesses would move there, pay no more tax, and destroy their old nation state.

This isn't just an idea. At least one country, Honduras, is planning on building a Charter City with laws from elsewhere. It changed its constitution in early 2011 to make it possible.

But just as some innovations are big enough to remap a country, others promise equally massive change, just on a much smaller scale—as we'll see in the next chapter.

12

3-D PRINTING

You need only open a trade publication to see that creative types get very excited by the prospect of "digital design" and "iterative design'—both twenty-first-century buzzwords. This is all very well, but if our computer systems can't actually produce physical objects, then they are going to be missing part of the whole industrial and creative world. For many years now we've been able to print out documents of course: we can print on paper or on cardboard, but those are just 2-D representations of the designs that we're making.

If we are doing real iterative design, if we are doing real testing, then we need to be able to make the objects themselves. Happily we now have a thing called 3-D printing. That's exactly what it sounds like: the ability to print objects rather than just sheets of paper.

3-D printers work by laying down the substance in layers, otherwise known as additive manufacturing. Most common 3-D printers print out a form of plastic but others can print out metal or even glass or ceramic, and as the print head moves from side to side, back and forth, it prints tiny thin layer over tiny thin layer. The object slowly builds up, and after a while you can take the completed piece out of the printer, use some small tools—files and so on—to remove the rougher edges, loosen up connections, and there you have it.

At the moment 3-D printers are quite expensive and a little unreliable. The cheapest one will cost you around about £500 but that can only make objects about the size of a tennis ball, and out of cheap plastic. The larger 3-D printers are still extraordinarily expensive and can still only make things out of one material. It is early days yet—but despite this 3-D printers are still incredibly useful, even when they are only capable of making small and simple parts.

Those parts can be incredibly complicated shapes—the sort of the shapes you get from messing around with your 3-D modelling software on your computer. Parts for your bicycle for example, or small parts that connect to larger parts: hinges, connectors and clips and so on. You can also make lots of beautiful artistic work, as we'll discuss in a moment. The point of a 3-D printer is that it enables you to make lots of different prototypes, making micro-manufacturing a real possibility. This means that you can design a prototype quickly, print it out, test it, see if it works, make the necessary changes to the design and print out another one, test that, and so on. By owning their own 3-D printer, a designer can try out lots of different prototypes without having to contract the work out to a major manufacturer and wait for each new design to come back.

Now that's very handy for a designer working today, but the really interesting thing about 3-D printing is the implications it has for the future. The first one is the science-fiction dream of 3-D printers: that one day every household will have one in the basement and whenever you need a new object you will be able to print it out. Break a glass, for example, and you can go down to the basement, load up the plans for the lost item from the Internet, and print out a new one. If you want some new cutlery, or you require a new button for your blouse or a buckle for your belt, you won't have to go to the store or order something online. Instead you'll just pop downstairs and print it out in the basement.

However, while that might sound appealing, these Star-Trek-style replicators would cause all sorts of additional problems. The first is copyright, and goes like this: given the fact that we all have very powerful computers in our pockets with cameras built into them, we can imagine the day when we have a 3-D scanner in our pocket. Already it is easy to envisage an app for your smartphone that would enable you to take pictures of an object from lots of different angles and then stitch the pictures together to create a 3-D model to feed into your 3-D printer. This means that if, say, you're at a restaurant and using a wine glass you particularly like the shape of, you could take a few discreet pictures of it with your phone, do a little processing and email it back home, so your new glassware is coming out of the printer when you get back from your meal. For objects with no inner workings—art works, pottery, jewelry and so on—the prospect of a handheld scanner application and a 3-D printer spells the same changes to the industry as the MP3 did to the music industry. Objects in the modern world are merely printouts of digital plans, and so the current idea that you can only get an object from its original manufacturer is destined to fade once the ability to do the printing out is more widespread.

The way our society starts to think about 3-D printing plans will be crucial in the coming years. Copyright aside, 3-D printers could be used to print things that we don't want people to have. Guns, for example, are heavily controlled in Europe, but there are plans available online for parts of weapons. A 3-D model of the magazine for an AK-47 assault rifle can be downloaded for free from the Thingiverse website today, for example. Inevitably, progress will not stop there, but is very difficult to consider how such technology might be regulated and in the end there can be no censorship regime for 3-D printing plans. If we can imagine it, and draw it on a screen, we will soon be able to print it out.

But that day is still yet to come. For now, we're stuck with on- and off-line shopping—and with advertisers, who are doing their darnedest to capture our attention, any way they can.

13

ATTENTION ECONOMY

Life used to be simpler, with far fewer choices. Take entertainment as an example. Total up all of the available entertainments on a wet Sunday in a small town, and in the mid-1980s there might be less than fifty things you could be doing. You would have only a handful of television channels available, perhaps one local cinema, a few radio stations, and not many books or magazines or newspapers to choose from. There wasn't much competition for your leisure time, especially if the passive consumption of media was what you were looking for.

Today, however, the Internet brings us the world. Not only are the choices available for the old media so much greater—pretty much every radio station in the developed world is available online, along with hundreds of television programmes and movies—but there are whole new genres and activities to choose between. Our time is claimed by everything from newspapers to social media to multiplayer online games. The amount of demand for your attention has increased massively over the past twenty years, but the available supply hasn't changed at all. There are still only so many minutes in the day you can spend doing things, or thinking about things, or considering buying some specific stuff.

For media companies and advertisers this is deeply worrying, and so the race is on to find new and interesting ways to attract your attention. A good deal of the innovation in the media world has been for precisely this reason. Social media is considered an enormously valuable business simply because it captures so much of our attention. If we're spending a lot of time on Facebook, say, and not watching television, then advertisers are going to move their spending over to the online platform. Selling your attention is how Facebook—indeed any media outlet—makes its money.

Taken too far, the metaphor of the Attention Economy might well sound depressing. It reduces the average person to a simple machine for consuming stuff—albeit media content rather than physical goods. Although with advertising that's simply an interim measure before you do your duty and buy the things your attention is being driven towards. However, let us not despair of the materialistic nature of society. Your position within the Attention Economy is rather more complicated than you might think—as a supplier of attention, you are increasingly fêted by content producers, whose demands would like to be met. As supply remains the same, but demand increases, the price goes up: the content gets better. If the price being offered by the person wanting to buy your attention isn't good enough, you won't buy anything and they'll die.

The Attention Economy, or at least the conditions that have lead to its importance, is one of the reasons why many old media organizations are dealing badly with the Internet. Their own concepts of the competition they face are still based on the pre-Internet world. A newspaper will compare itself with other newspapers, a television station with other television stations. But neither comparison helps today. Take news: pre-Internet, I could wait for an hourly radio bulletin, or wait longer for the evening television news, or wait longer still, until the next morning, and read the printed newspaper. Each had its own place and probably didn't compete with the other. Today, however, each competes for my attention with the Internet, and not only in the moment, but over a rolling period: I'm far less likely to buy tomorrow's printed paper if I've been reading news online all day. This also, of course, assumes that I'm really interested in news, and not just in passing the time, and that is by no means clear. So today's newspaper is competing within the Attention Economy not only with all of the rolling news coverage on the Internet the day before, but also with all of the Internet sites showing pictures of kittens, with all of the computer games I can access, and so on. Worse still, the four or five quality newspapers in the U.K., where I live, are also competing with all of the other quality English language papers around the world. On the Internet, the *New York Times* is just as easy to access as the *Guardian*.

This is not to say that the Attention Economy only goes in one direction. While the media industry and business might be trying to get you to spend your attention, you might find that you're better off investing it: spending it wisely on things which might be improving in the future. As individuals we need to pay attention to what we pay attention to, and we need to make sure that the attention we're paying is the best possible attention that we can apply to a particular subject.

You can only spend your attention once, so it's worth spending it wisely. It is this genuine scarcity that makes it interesting. Anything you pay attention to is preventing you from paying attention to something else.

In many ways this is a fundamental shift in the human condition. We have gone from people who were attention rich but information poor, to people who are information rich but attention poor. There's very little that we can't find out, given perseverance with a search engine: but few have the time or inclination. This could be seen as an illness, or at least as mildly dysfunctional. In fact, as we'll see in the next chapter, many people are indeed trying to shed everything that claims their attention, in a bid for sanity.

14

TECHNOMADISM

Back in the last century, futurist pundits would talk about telecommuting or telecottaging: the idea that using modern communications technologies we wouldn't have to go into the office. We could just work from home in our pyjamas, using email and video conferencing to get our work done without having to bother with a commute. It was a vision that was written about again and again from the Fifties onwards. But it has only really been in the past five or six years that such work style has been possible. It is only in the past few years that video conferencing has gotten good enough and cheap enough, and collaborative software and email has become easy enough to use and powerful enough, for people to feel liberated from their office.

The realization that in fact you get much more work done at home in your pyjamas than you can in the office when you're being distracted by your colleagues has come at the same time as people started to realize that they can carry their entire music collection around in their pocket or access any form of content or entertainment they are interested in simply through a web browser. This, plus the advent of social networks, means that we can push our work life and our leisure time up into the Cloud.

Once everything is on the Internet, of course, distance ceases to matter. If you're telecommuting—if you're video conferencing into your office—it doesn't matter whether you're at home or sat on the beach a thousand miles away. As long as you have a machine that can access your data you are effectively still part of the modern world. For a small but ever-increasing class of work, the only thing you need to be able to do is access your data, and therefore an ever-growing number of people are realizing that they don't actually have to stay at home at all. They can travel the world and still make a living as web designers or journalists or writers or some other form of knowledge worker perhaps just as easily as if they stayed at home and paid their rent.

These technomads have taken the Cloud and telecommuting to its logical conclusion and have gone walkabout, while still remaining connected to their origins.

The fashion for technomadism, or at least the aspiration for it, is also driven by another social movement that is occurring in the second decade of the century: neo-minimalist or simplicity. If you are a young coder who is planning on travelling the world and working from your laptop as you go, you are faced with one very large question: what are you going to do with all of your stuff? Whereas in the 1990s and 2000s people would have taken their spare stuff to storage, only to come back to it a few years later, the fashion in the 2010s is to give all that stuff away. In 2010 the writer Dave Bruno took the Internet and the American press by storm with a blog post called the "100 Thing Challenge," in which he set out to pare his possessions down to a mere hundred objects. Many have followed in his footsteps, with the challenge spawning many other websites dedicated to a life of simplicity and minimalism.

The movement has come under a great deal of criticism: getting down to a hundred objects is still owning a good deal more objects than the vast majority of the world, many of whose citizens don't see a hundred objects as a lifestyle choice or a route to Zen happiness, and it doesn't escape notice that most people's hundred objects includes a laptop capable of holding thousands of albums worth of music. Today that one single object can take the place of tens of thousands of objects from a few decades ago. I myself have undergone this process, sold all my books and given away all my CDs, and if I count all my cutlery as one object and my underwear collection as a second object then I'm well within the hundred objects threshold. Personally, I find having less stuff around me helps with my personal Attention Economy. It's a way of being able to concentrate, when otherwise I might find myself distracted by clutter. This is perhaps an extreme reaction to a wandering mind, but it's not nearly as radical as our next idea: academic doping.

15

ACADEMIC DOPING

It will come as no surprise to anybody reading this book that the world is increasingly pressured for knowledge workers. The demands on focus, concentration, mental stamina, intelligence and dedication to the work on the screen in front of you grow larger and more onerous every day. There probably isn't a single person reading this chapter who doesn't feel under pressure to have better ideas, to concentrate harder, to work for longer. This is especially true in academia, and so one of the trends we have been seeing over recent years is the rise of academic doping.

Some athletes have, of course, been taking drugs to improve their performance for many years, but it is only in recent times that drugs have been available that seem to, at least in the short-term, increase intellectual performance. Drugs which are usually prescribed for mental conditions like ADHD or narcolepsy are being taken by patients who have neither of those conditions, but who want to improve their academic performance. I myself have experimented with the anti-narcolepsy drug Modafinil, which if taken at the right time of day will enable you to stay awake for at least twenty-four hours with seemingly no effect on your mental acuity. It's given to soldiers and military pilots for much the same reason. The anti-ADHD drugs Adderall and Ritalin are taken regularly in American universities, both by the students and their professors. Reports from as far back as 2005 and 2006 have high school students—fifteen-, sixteen-, seventeen-year-olds—being taken to the doctors by their parents for a prescription for Adderall in order to help them improve their grades.

Surveys in America show that one in four college students and one in ten middle school and high school students are taking some form of drug to help them with their school work.

This raises lots of interesting moral and ethical questions. Many people will ask, what's the big deal? After all, people take vitamin pills and go to the gym and read improving books, and all of these things can help you enhance your performance in some way. What difference does it make that rather than reading an improving book or meditating you're taking some pharmacological aid to help you think more clearly or for longer periods of time?

One argument against cognitive improvement drugs is that they are unfair in an egalitarian sense. In other words, some people will be able to afford them and others won't. Remember, the advantages of these drugs can actually be quite marked: after all, if I can get an extra couple of hours of hard thinking in during the day, then that is equivalent to another day in the week—perhaps two extra days a week—where I can get some effective work done. If that sort of mental efficiency is only available to people who can afford it, we will see a growing inequality in our society. Competition between people could in effect simply become a competition between the drugs that they are able to afford.

However, this argument doesn't really work. There are many ways to get smarter and none of them are equally distributed around society—not least schooling. Education itself is very unevenly distributed but nobody thinks of banning it for that reason.

The second worry about academic doping is the effect it may have on our health in the long run. There have been very few long-term studies on the effects of these drugs on people who are using them for off-label reasons—i.e. taking the drugs without having the condition that they were originally designed to treat. This of course is a huge risk, and many of the early studies are starting to show that people who use them in this way find themselves at a greater risk of things like depression. They could also become addicted—if not physically then certainly psychologically, as the ability to go without sleep and get an awful lot more done can become very important to peoples' everyday lives.

A third worry is how people get these drugs in the first place. Most of the drugs are prescription-only in many countries, and so anyone using them for off-label purposes has to either con their doctor or buy them from somebody who has a genuine prescription. In some countries, and for some drugs, a prescription isn't necessary. In the United Kingdom, for example, Modafinil, the anti-narcolepsy drug which, if taken by someone who does not suffer from narcolepsy, can keep you awake and alert for a long period of time, doesn't require a prescription and is easily available to buy over the Internet. It's completely illegal to do that in the United States, whereas Adderall is much more easily available there than it is in the U.K. Such availability is illegal. These varying degrees of legality and availability mean that students may run the risk of becoming drug dealers simply to get their essays done. Some people, parents especially, might find this morally and ethically unacceptable.

Over the next ten or twenty years, the U.K., Europe and North America are all facing an ageing population. One of the biggest worries that this brings is the proliferation of age-related illnesses, and specifically age-related mental illnesses—Alzheimer's, for example. Because of this we are able to foresee a whole new range of anti-Alzheimer's cognitive-enhancing drugs available on the market, and thus more off-label use and more and more academic doping. Just as it was once said that a mathematician was a machine for turning coffee into theorems, and no good newspaper headline has ever been written without the aid of a cigarette, we may soon find ourselves in a culture where the idea of thinking for a living without taking these drugs is itself unthinkable.

All this talk of drug-taking comes from a background of increased knowledge of how we think about ourselves. Over the past few years we've begun to understand all of the different ways of helping ourselves to create a more productive environment. Meditation, the flow state, personal productivity systems have all come together to create an environment in which we are able to think better, for longer, more clearly and more imaginatively, even without the aid of drugs. An athlete can't just take steroids and become a champion automatically. They still have to go to the gym, and train, and eat the right things, and do the right exercises. And so it is for academics and knowledge workers. Yes, there are drugs available, but there are also other techniques that enable us to become clearer thinkers and to exercise our imagination in better ways. And it is those techniques that perhaps we should be concentrating on.

Especially since, as we'll discover in the next chapter, Big Brother may well be watching us.

16

OPEN GOVERNMENT DATA

Our governments know a lot about us. They have our data. They know where we live, they know how much we earn, they know about our health and about our education. They also produce a lot of data of their own. The locations of all government offices, where the schools are, where the hospitals are. How good the teachers are in those schools; how good the doctors are in those hospitals. How efficient the rubbish collection is. What days the rubbish is collected.

In sum, a government produces an awful lot of data. Data that describes the services that the government offers and also describes how good those services are. There is also data on how people use their services and who those people are. This is information that we ourselves as citizens actually own. After all, we pay for government services and we're the people who use them. So there's a growing movement around and within governments to release that data into the public domain, so that we can take it and use it ourselves.

There are many reasons why you might want to use government data. First, it introduces transparency. In a democracy it is very important that the citizen knows what the government is doing. It is not enough just to take the politicians at their word. There are many ways of measuring how a government is doing, of course, but the sort of data you can get in terms of school grades or hospital waiting times is a very good way of measuring the efficacy of education and health policies. So by releasing government data, the government allows the citizen to see how the government itself is doing. This is incredibly important, but it isn't the only thing. Transparency isn't just about access to data. It also requires us to be able to use that data, to share it with each other, and to be able to roll it into our own applications, combine it with other data and produce our own conclusions. Open government data isn't just a matter of releasing a database or spreadsheet. It is releasing the data in a way that allows computer programmers to combine it with other data and give us some interesting insights.

That brings us to the second reason for a government to release its data. It creates value. It provides an opportunity for entrepreneurs to take that data and produce goods and services from it. Let's take a restaurant review site. Although it creates its own reviews, sending reviewers to go to the restaurant to see how the food is, a site could also make use of the government's own health inspectors' reports. By taking the government data that says how clean a restaurant is and applying it to the review of how good the food is, they can create an even better service that both adds commercial value to that business and makes the government health inspections more potent.

It is a way of getting data from the government that we have already paid for and, by adding to our businesses, of producing more wealth for the nation.

The third reason for a government opening up its data is that it creates an atmosphere of participation. In the same way as data coming from the sensors embedded within Smart Cities, as we will see later, enables you to choose which way you want to go home in the evening in order to avoid traffic, having real-time data from government services enables politicians and citizens to come together to create better policies, almost in real time. We can no longer wait for four or five years for an election before we make decisions based on things that are happening today. Opening up the data that the government is producing allows us to interrogate exactly what is happening in the nation. This works on a national level, on a city level and on a supranational level, across the EU, or across North America, or across the world. There have been a long series of campaigns internationally to get governments to open up their data over the past few years and these have been increasingly successful. The British government, the American government and many European, Australasian and Asian governments have all opened up datasets. Currently there are more than 700,000 government datasets available to anybody to use around the world. These include things like crime statistics by local area, as well as the schools and hospitals data we've already mentioned.

The most famous applications to come out of this open data are the Chicago crime maps. Released in 2008, they are maps of the city of Chicago showing where the crime is, street by street. This is the sort of government data that enables you to make decisions about your own life: where to live and where to send your kids to school.

In the U.K., the organization Rewired State brings together web developers and government departments to produce new and interesting applications and projects. There is even an annual Young Rewired State competition to encourage under-eighteens to produce their own applications based on data from the U.K. government. A good example of this was the application that won a few years ago, where a sixteen-year-old had taken data about the locations of schools from the Ministry of Education, crime data and local transport data, and added them together to create an application that would plot a route from your home to your school that avoided areas you might be mugged.

And now from urban space to outer space . . .

17

AMATEUR SPACE TRAVEL

With the last space shuttle mission ending on 21 July 2011, we could be forgiven for thinking it was the beginning of the end of space travel. The retirement of the first reusable spacecraft means that it's impossible for the United States of America to put astronauts in space—at least without the help of the Russians. It would seem that after sixty years of government-supported spaceflight, all that sort of adventure is coming to an end. Even the astronauts currently aboard the International Space Station will soon be headed back down to Earth. Soon official government agencies won't have the capability to put a man into space at all. So what are we going to do? Well, the second decade of the twenty-first century is to see the birth of amateur space travel. Not simply private companies putting satellites into orbit—which has been happening for quite a few years—but putting astronauts into space, scientists onto space stations and tourists into Low Earth Orbit to look at the view.

The next few years will see the beginnings of spaceflight being dominated by a handful of organizations, all privately owned. Some, like the Virgin Galactic operation, will take tourists up to the very edge of space, just far enough for them to experience weightlessness for a few minutes, before returning them back to Earth around £100,000 the poorer. Other teams, like Space X, are building rockets to launch satellites into Low Earth Orbit, and even bigger ones to be able to take astronauts and supplies up to space stations. Others still are dedicating themselves to building space stations that can be delivered by these rockets, and have scientists and astronauts ready to live upon them.

Bigelow Aerospace is a private company based just outside of Las Vegas, which even today is advertising space stations to lease. The price includes the training of the astronauts. Their website says that, "whether you're a sovereign nation developing an astronaut programme, a corporation interested in microgravity research, or an individual with the desire to experience space, we can help you achieve your goals."

The very possibility of private spaceflight has come about because of a competition. The X Prize, which was started in 1996 by Dr. Peter Diamandis, was a competition with a prize of $10 million for the first company to build and launch a spacecraft capable of carrying three people 100 kilometers above the Earth's surface twice within two weeks. That first prize was won in 2004 on the very day of the forty-seventh anniversary of the Sputnik launch, by a project paid for by the co-founder of Microsoft, Paul Allen. His ship, called SpaceShipOne, won the prize and was awarded the $10 million. However, the whole flight had actually cost Allen over $100 million to complete. Nevertheless, the success of this X Prize has led to many more, including the Google Lunar X Prize, which made $30 million in prizes available to the first privately funded teams that were able to safely land a robot on the surface of the moon, drive it 500 metres, and then send video, images and data back to Earth. That $30 million probably isn't going to pay for the entire mission, or even a fraction of it. But there does seem a curious part of human psychology that reacts extremely well to this sort of competition, and there are currently twenty-six teams competing for the price. They have until the end of 2015 to get there, meet the prize objectives and win the money.

Space exploration isn't restricted to governments and very rich individuals running their own private companies, however. We are now beginning to see forms of satellite which we could soon pay for on a standard household credit card. Universities and private companies can buy "CubeSats," which are kits to build satellites which are 10cm x 10cm x 10cm, and as long as it reaches certain weight requirements and is strong enough to withhold the vacuum of space, you can fill your CubeSat with any form of gadget and sensor you like.

You can order a basic kit off the Internet for about $8,000, plus postage and packing. Because you can fit an awful lot of CubeSats into the space taken up by an ordinary satellite, it is becoming ever easier to get your satellite launched into space. The cost of launching satellites is measured solely in terms of their weight. In 2007 it was $50,000 per kilo, but over the next decade this price should drop as all of the X-Prize-inspired private companies start launching their own private rockets into Low Earth Orbit.

Still, if the $300,000 it cost Colombia to launch its first ever spacecraft is a little rich for your blood, there is an alternative: an even smaller type of spacecraft called "the Sprite."

This prototype system is a couple of chips on a circuit board the size of a postage stamp which, if put into orbit, is able to make radio contact with the Earth. Sprites are so tiny you could fit hundreds inside a single CubeSat—which means the launch costs could be incredibly low. There is a crowdfunded project on the Internet, as I write, that is offering the launch of a Sprite into space, transmitting your initials in Morse code for $300.

A satellite beeping your initials in Morse code isn't that useful, admittedly, but as with everything else mentioned this book, Moore's Law is at work here. A satellite that can do very little today might be able to do an awful lot in ten years' time, just in time for the private space companies to be around to launch it into orbit.

These aren't just playthings. Real science can be done with them. There is also an ever-increasing demand for communications satellites and remote-sensing satellites that can take a look downwards at the Earth. These sometimes have surprising uses. The bank UBS used satellite imagery of the shops belonging to Walmart to predict their quarterly sales results. They were counting the number of cars in the car parks, and extrapolating from there.

Satellites aren't simply for First World countries either. The developing world is starting its own space programmes. Nigeria, for example, spends $50 million every year on its space programme, and has already launched two satellites. Asian countries are even more advanced: India has been launching satellites for many years, as has Japan and China. The Chinese have declared their intention to set foot on the moon by 2024 and then go from there to Mars sometime between 2040 and 2060.

Exciting stuff. But, as we shall see in the next chapter, there are also adventures we can share from the comfort of our own home.

18

GROUP BUYING

If I were pushed to name the single most significant thing that the Internet allows us to do, I would probably say the forming of groups. Pre-Internet, if you wanted to assemble a collection of like-minded people for the purpose of protesting or socializing or trading, you had to arrange for them to be in the same place at the same time. Now we can be elsewhere and else-when and still gather to stage a revolution, fund an art film, gossip or sell each other stuff. The ease of group-forming is at the heart of how the Internet affects our lives.

One of the new business models that can only exist because of the arrival of our digital world is group buying, as practised by companies such as Groupon. These companies allow you to harness the buying power of a large mass of consumers, without the hassle of having to put together your own collection of people who all want to buy the same thing at the same time.

Groupon and their ilk hunt out businesses that can make hugely discounted offers to consumers, so long as they are guaranteed a certain volume of sales. Once the required number of buyers are ready to part with their cash, the deal is on and the consumers reap the benefit of Internet-powered group-forming, to the tune of 60% off a tailored suit or 70% off a day at a spa. Although why anyone would want a fish pedicure—even at a deep discount—remains a mystery to me.

Then there's Kickstarter, which is less about tempting offers and more about creative entrepreneurship. It uses exactly the same principle as Groupon—connecting producers and purchasers—but with the aim of building a funding platform for brand-new products, social enterprises or artistic projects. If you want to make a documentary about the sex lives of sloths, you can post your proposal on Kickstarter so that anyone who is intrigued by the sheer unlikeliness, or the slow sensuousness, of sloth procreation can pledge money to send you out to the jungles of South America. Once the funding target has been reached—and only then—money is released and the project is required to deliver within a certain time frame or forfeit the cash. With tiered levels of funding, the backers might receive anything from a DVD of the film, to a DVD plus an invitation to its premiere, to a DVD, invitation and a role on the filming expedition. It's a twenty-first-century system of patronage which is completely dependent on the unique power of the Internet to bring people together.

The benefits for the producers are clear: you get your film made without losing any creative control to a single big backer. Or if you are launching a chocolate-flavoured lipstick, say, you get to test the viability of your idea with no risk attached, and raise capital at the same time.

The benefits for the purchasers are more intriguing and more specific to the new digital world. If you pledge £15 to the sloth film, you are paying for a great deal more than the DVD you'll eventually receive in the post. Once you've committed the money, you'll be updated at every stage on the project's progress. You can access blog posts on the team's week-long hunt for the sloths, share the excitement of trekking through the rainforest, the exhilaration of filming sequences that have been years in the planning. The DVD, when you eventually receive it, will have taken on many Spime-like attributes: it will be an object about which you know a vast amount (see Chapter 45 for more on the wonderful world of Spimes). And your £15 will have bought you access to a unique creative vision and the thrill of making something you value happen. That's the kind of heady experience that used to be the preserve of Popes commissioning Renaissance artists, or venture capitalists backing the latest social network. Never before has it been possible for an individual with limited means to be an enabler of creative vision on this scale. That's the power of group buying.

19

STUXNET AND SONS

Sadly, not all technological innovation is so benign. We have now had computer viruses for more than thirty years. The first one was written in 1972. The Morris Worm was the first to spread over the Internet in large numbers: it was written in 1988. Since then, thousands of different computer viruses have been released, and while many have been destructive, and some have been very virulent indeed, they have never been considered a form of warfare. Nor have they been specifically sophisticated to a degree that has caused international alarm. In 2010 that all changed. The first computer virus to be seen to be an act of international sabotage, and perhaps a new form of warfare, was captured in the wild.

In June 2010, a Belarusian security company called VirusBlokAda discovered a virus that was exceptionally sophisticated. Analysis by the anti-virus community showed that it was able to gain control of infected machines by the use of four Zero-Day Exploits inside Microsoft Windows. A Zero-Day Exploit is a previously unknown security hole; and in an operating system as studied as Windows, if you're smart enough to find even one the monetary rewards can be enormous. To use four of them in one virus, then, was almost inconceivable.

The second unusual thing was that the virus was specifically designed to look for a type of industrial machinery called a SCADA (Supervisory Control And Data Acquisition) controller. These devices control industrial equipment in factories. This new piece of code, which the researchers called Stuxnet, was written to target a couple of very specific models of SCADA controllers, and to make them spin their machinery—gas centrifuges—at very specific speeds for set periods of time, while reporting back to the infected machine that all was well. It would not be, as the speeds Stuxnet would force would cause that machinery to eventually, and seemingly inexplicably, break.

This made no sense to the researchers, until it became apparent just who used those particular models of controller. Stuxnet's target, it would seem, was the uranium enrichment plants of the Iranian nuclear programme. There, in a factory called Natanz, UN weapons inspectors had been mystified as to why 1,000 of the 10,000 centrifuges had all broken at precisely the same time, just as the Stuxnet code was first released.

Stuxnet was written specifically to target the Iranian nuclear programme—if not to stop it, then to slow it down considerably. It was only by accident that the virus was discovered in the wild in another country, deconstructed in public and its exploits fixed, before it had gone on to do further damage.

No one knows who wrote the Stuxnet code, nor how it came to be inside the Iranian plant, but the sophistication of the attack makes it most likely to have been a nation state, rather than a group of amateurs. Evidence suggests it was either the United States or Israel, or perhaps both, who did the work. No one will say. It is, of course, irrelevant. The point of Stuxnet is that it shows that with resources and will, specific industrial machinery can be attacked with computer code, not just with bombs. Stuxnet was programmed to disable itself if it found it was infecting a machine that didn't control those very specific gas centrifuges. It was written, it seems, for that one target.

That shouldn't let you relax. In October 2011, the anti-virus community announced the discovery of a virus they called Duqu. Duqu shares some of the same code as Stuxnet, which suggests that it was either written by the same people or that the Stuxnet code is being sold somewhere. It doesn't target nuclear processing plants, but rather is aimed at specific corporations. Duqu installs itself, gathers system information about the machine it is on, and the network it is connected to, and records all of the keystrokes made on that infected machine. All of this data is then sent back to a command and control center somewhere on the Internet.

In the version of Duqu found in October 2011, that center was an address hosted in India, and before it was shut down it seems to have sent out three other payloads to Duqu-infected machines. These payloads looked for more information, including the time zone of the machine, the names of their connected drives, and a screenshot of whatever the infected machine was displaying at the time.

All of this information would go a long way to help make a more lethal child of Stuxnet. Again, it seems that someone—and at time of writing we have no idea who—is releasing custom-made computer viruses onto the Internet that target both military and civilian networks. Given Moore's Law, and the inevitability of the Stuxnet code being leaked into the public domain, we can imagine a good deal more of this in the coming years.

20

SPATIAL FIX

Super-viruses aside, there's still plenty to keep Joe Bloggs awake at night. We are living in a decade of huge economic change. The financial crises of the last decade have created an economic situation worse than has been seen for many generations. That, combined with the effects of the Internet as discussed in this book, come together to create an environment completely different to that of our parents' or grandparents' generation.

Major upheavals in our economic life are always accompanied by the introduction of a new way of living. The move to the suburbs after World War II is a good example of this. An economy based on light manufacturing and blue-collar manufacturing jobs enabled people to live in the suburbs and travel by car to their factories and places of work.

Today, however, knowledge work—which in the West still seems to be the dominant economic force of the twenty-first century—doesn't require large spaces. Instead, it seems to work better when people are living in high-density cities. The academic and writer

Richard Florida has been very influential over the past ten years with his writings about Creative Cities. His theory is that cities that are full of creative industries and creative people naturally gather certain characteristics. The Creative Class, as he calls them, have a specific lifestyle which is facilitated by urban living. They like, he claims, street cafes and galleries and bistros. The more people in the Creative Class, the more these cafes and museums and galleries thrive, which in turn will attract more creative people to the city. The theory goes, therefore, that if you want your city to prosper and to attract more creative people, then you should concentrate initially on projects which attract this sort of lifestyle rather than, for example, building a large stadium. His ideas have been very popular. You can see them in many small and medium-sized towns that have created their own creative quarter or special areas for artists. They're hoping to attract other creative people and bring prosperity back to a community that has in the past relied entirely on manufacturing or other businesses—businesses which have now moved to China and the developing world.

Today, with the economy worse than ever, many urbanists and other supporters of Richard Florida's work have called for an assessment of the types of lifestyle that people in the West desire. Whereas in the Seventies, Eighties and Nineties perhaps the most desired lifestyle was a suburban one, in the second decade of the twenty-first century it's all about the urban dream.

In a knowledge economy, the most important thing is ideas. And, according to the research of Geoffrey West of the Santa Fe Institute, cities with more people per square metre produce more ideas and more money per person. Encourage people to move away from the suburbs and gather in cities where the Creative Class lifestyle is promoted and, the theory goes, we all become happier and wealthier as a result.

As the years go on, we may find ourselves living evermore in the center of cities that we had previously abandoned. This change in lifestyle is called a Spatial Fix. But what does it mean practically? Well, people will be more mobile, there'll be less property ownership and more access to services and amenities. By giving up the suburban dream we may find ourselves less rooted, but happier.

Many people already understand that we will be required to be more fluid in our job desires and skills over the next few years, rather than being set in one place during one thing. This is not without its downsides, of course. We will require a whole new set of social constructs if we can no longer rely on the community of geography. People who live in the suburbs today, or in small, unsustainable towns, will find their lives under a great deal of pressure in the next few years. The desirability of moving to a big successful city because of the changes in the global economy means that whole areas, and indeed whole populations, may be left behind, ignored by the international networks of globalization.

21

MEMES

If we live in a world where ideas and information are more important than physical objects, animals or people, then it's a very good idea to understand how ideas actually work. Not so much the specifics of the ideas—it would be very difficult to create a general rule for making a good idea that wasn't so broad as to be useless. But rather, we need to understand how ideas travel; how they spread from person to person, company to company, or country to country. We already have lots of words for this process. There's education and knowledge sharing. There's fashion and gossip. There's story-telling, and explaining, and preaching, and many others. But all of these words at their heart come to mean the same thing: memes.

Pronounced to rhyme with dreams, memes are the most basic form of idea, the single indivisible unit of culture. They're the thing that moves from person to person as we tell jokes, or teach songs, or pass on etiquette. There are at least sixty-four of them in this book, and hundreds more in every newspaper, conversation or any other human interaction.

Richard Dawkins, the scientist who came up with the meme of memes in the first place, considered that a meme was a piece of culture that replicated. You might have an original idea, for example, which becomes a meme when it is replicated in the mind of another, when you tell it to them. But because people aren't always very accurate when they replicate an idea, memes tend to develop different versions of themselves. Some of these memes will be stronger, and more popular, and so replicate more often. Others will be weaker, and not as good. They don't replicate quite as fast. A funny joke gets told more often than an unfunny one. A funnier joke goes even further still.

And this is the key point: in this aspect, memes are subject to natural selection. In a curious way, memes are alive. Like anything else in the living world, memes are subject to the rules of Darwinism and epidemiology. The fittest ones live on: they breed and they spread and they evolve. Unfit memes die, literally forgotten.

When we talk of the replication of memes, this includes things like writing, and not just direct, face-to-face explanation of an idea. Writing is, after all, just time-shifted and place-shifted imitation. I'm typing these words months, if not years, before you read them, but I am hopefully spreading a meme as I do so, reaching through time into your brain.

This is one reason why the Internet has been such a powerful force on culture and society in the past twenty years. The more things we read, the more sources we sample, the more information we take in, the more likely we are to be infected by a meme. The Internet is highly contagious, and not just alive with memes you've encountered before, but with whole new ones you don't have immunity to yet. Hang out in a new part of the Internet, and you're as likely to catch a new idea as someone in an airport is likely to catch a cold.

It's a powerful metaphor, because it allows us to use the tools and ideas we developed for explaining biological systems, and specifically the spread of diseases, to explain how pieces of culture travel through populations. Those pieces of culture can be anything from a perception that a particular cut of dress is fashionable this year, to the ideas of global jihad, or right-wing libertarianism, or that a particular brand is cool.

To this end, memes are most invoked in the non-academic world by advertising executives. The very point of advertising is to infect you with the meme that either the product in hand is good and worthy of buying immediately, or that a particular brand has certain attractive values that will rub off on you should you buy any of its stuff. The concepts of "luxury" or "in fashion" are themselves simply memes, not measurements of anything real.

Memetic advertising reaches its extreme with the concept of the "viral campaign," where a meme is engineered so successfully that it spreads far beyond the original starting point. Funny videos on YouTube, for example, can be such pleasing memes as to compel the infected to infect others by sending the link on. We have all done it.

The meme idea goes further than merely giving us a way to sell more stuff. It points out that ideas aren't successful because they're fundamentally important, but because they are designed, or have evolved, to be so. Memes are successful not because the idea they contain is good, but because the meme itself is highly infectious.

This explains why societies can be frustratingly obsessed with the trivial—reality television, say—and disinterested in the important—global warming, for example. The "trivial" is simply a more powerful meme. As Noël Coward once said, it is "extraordinary how potent cheap music is." When a meme evolves to the heights of infectiousness, we get a fad, or hysteria, or another such passing madness. These are infinitely more compelling than a dull meme, languishing in the background, no matter how serious.

Memes evolve as they spread, making small changes to each generation, with only the fittest surviving for long. The best memes are therefore the most evolved, and the most evolved are the ones that have spread the furthest. We can say, then, that a good rule of thumb for getting a better idea is to let it spread and evolve freely in the minds of other men.

22

CROWDSOURCING

We've already seen that facilitating people's desire to gather into groups in a way that is neither space nor time dependent is one of the Internet's pre-eminent characteristics. The ease with which communities of interest are generated catalyses political activism and leads to a million million-man niches, to crowdfunding, and group buying. It also enables crowdsourcing: like outsourcing, but channelling work to disparate groups recruited online.

The question of how to harness the power of the Internet's potentially enormous groups of people has been exercising everyone, from activists to marketers, for years. It has not escaped the attention of the business world that asking a lot of people to each do a tiny amount of work could enable you to complete certain large-scale tasks in an extremely cost-effective and speedy fashion. You might, if you were very clever in your use of gamification, be able to get people to do your work without paying them anything. If you were even more cunning and could disguise your work as something that was essential to them rather than to you, you could get them to complete it without even realising they were working at all.

In some ways, the canonical crowdsource operation is Wikipedia, an idea of genius simplicity that has established itself as one of the biggest authorities on the Internet. Wikipedia shouldn't have worked in theory but absolutely did in practice, and showed that people could and would work collaboratively when it was fun, and in the service of an ideal. Wikipedia was not so much an enterprise's commercial product as a manifestation of the do-it-yourself-and-share-it idealism of the Internet. But its success certainly focussed the minds of business people.

The standard business application of crowdsourcing is available at Amazon's Mechanical Turk marketplace. If you have a task that you can split into small components with fees to match, you can post it to Mechanical Turk where a global workforce will complete it for you. An example of the sort of job one might find: if you have 8,000 photos of used cars for sale that need tagging either left-hand or right-hand drive before they are uploaded to your site, it will be cheaper, faster and more efficient to crowdsource that task than to recruit a team of temps to spend weeks staring at the photos, getting bored, taking long breaks, falling asleep and giving up. And you need human intelligence for this task because artificial intelligence lacks the necessary powers of intuition to analyse the often fuzzy, wonky, unhelpful photos provided by sellers. If you split that task into thousands of tiny packets, you can get it completed in less than twenty-four hours for a relatively nominal sum. If you are concerned about inaccuracy, you can run the photos through Mechanical Turk twice. It will only take another twenty-four hours and the cost will still be tiny.

One retail website discovered that poorly spelt and punctuated customer reviews were damaging sales, even when the reviews were positive. So it posted the proofread of its entire database of customer reviews to Mechanical Turk. Twice. Within days the spelling and punctuation had been cleaned up by a vast team of anonymous and unconnected workers, working for pennies (relatively) and sales were rising dramatically.

It's not all about straightforward commercial applications. Crowdsourcing has also been used by organisations such as the U.S. Navy to explore new techniques for the deployment of its forces. The first such exercise was launched in May 2011 to develop new strategies for combating Somali pirates. The Navy routinely holds conventional war games to test out new ideas, but those ideas are generated by the war game participants: usually senior officers. The problem is that senior officers tend not to be the best people for dreaming up (as opposed to implementing a realistic version of) ingenious ways to take on new enemies. So the Navy announced that it was recruiting participants to play in MMOWGLI (Massive Multiplayer Online War Game Leveraging the Internet), threw in a huge dose of what's known in the business as "gamification" (more on this later) and observed the results of a thousand civilian and military volunteers playing out how to defend against pirate attack, rescue hostages and protect commercial shipping from disruption. The benefit for the Navy is obvious: they glean a host of insights that might otherwise have been unavailable. The costs of developing the gaming platform were immense, but it is sufficiently flexible to model other situations. And there will never be any shortage of people who would relish the challenge and the fun of gaming on behalf of their country's interests.

An even more ingenious use of crowdsourcing comes in the form of ReCaptcha technology. This was a development of the Captcha technology, the character-recognition system that asks you to type in the letters of a scrambled word in order to access a website. Captcha is a security device designed to catch out spam bots, but ReCaptcha is a commercial tool that uses human intelligence to resolve problems in digital scanning of old texts. For example, if you are trying to scan and archive a newspaper library online, you are relying on artificial intelligence to carry out the majority of that task. But there will always be some words that, because of faded ink or creases in paper, a computer cannot recognise. A native speaker, however, will be able to identify them intuitively with little difficulty. ReCaptcha posts a second word for you to type into a little box in order to access the site you want. It's the work of seconds to complete, and you are just focussed on placing your Christmas gift order or checking your email, so you think nothing of it. But in that moment, you and hundreds of thousands of other users have contributed your labour, for free, and without even realising it, to a commercial project. There is now so much ReCaptcha technology online that its users digitised twenty years' worth of the *New York Times* in a couple of months.

Crowdsourcing was (briefly) fashionable in political circles, for understandable reasons. Asking voters their opinions is something that politicians have always done, but this is a step beyond even the much-maligned focus group. Crowdsourcing policy documents for correction and feedback seemed like a good idea, until it was actually tried. It turns out that when a task is too complex, splitting it into small packets simply strips it of its context and results in either extremist or banal responses. Rather than a series of discrete tasks that get completed, the job becomes saturated in its multiple participants' views. There is also the attendant image problem: if you are a political party or a big corporation, you need and want to engage in dialogue with and listen to the views of your target constituencies. But that can very easily be perceived as lacking any credible and original ideas of your own. It is a paradox of the social-media-saturated Internet that the most successful digital product vendor, Apple, is the one that takes the least notice of anyone other than its own people.

There may be limits to what can be successfully crowdsourced, but the basic truth of the Internet's networked structure means that we will see more and more of it creeping into more areas of life.

23

HIGH-FREQUENCY ALGORITHMIC TRADING

Anger at bankers over movements in the stock market has been a common theme for the past few years. But in some ways that anger is misplaced. Most of the time it's actually computer systems which not only control the timing, amount and price of trades, but make their own decisions as to how to proceed. Indeed by 2011 more than 70% of all of the trades made on the U.S. stock market were made by automatic systems called Algos, rather than by people. Algos, short for algorithms, are sets of rules and artificial intelligences that monitor the market and decide, based on their programmed strategy, on what, how many and when to buy and sell. In other words, the stock market is mostly AIs running the financial world on their own.

The Algos belonging to major banks are closely held secrets. Their trading strategies would be useless if others knew what they were doing, as other banks would write other Algos to profit from them. While traditional investors might hold a stock for days or months, an Algo might only hold theirs for seconds or less. High-Frequency Trading Algos can make trades every millisecond or faster, taking tiny profits each time, but making thousands of such trades every day.

This sort of trading depends almost entirely on the speed that the Algo can get the data. If my Algo knows a price before yours does, for example, it creates an opportunity for a trade between the two—and I'll profit from it. Because of this, a new geography is starting to form. The ultimate limit to the speed of the price getting to my Algo is the distance between it and the systems it needs to talk to. For every hundred miles, there's an additional one millisecond pause. Even hundreds of metres can make a difference, and so special data-centers are being built as close as possible to the servers that run the major exchanges, by the banks that want an advantage. Some of these "proximity hosting" data-centers are even built with circular server racks, as opposed to the usual rows of cabinets, to minimise the length of cabling between machines, and save vital tiny fractions of a second. Other companies are laying their own transatlantic fiber-optic cabling, to minimise the time needed to send a signal from London to New York, say.

The future of this trend will not be a surprise to readers of this book. With the increase in desktop computing power brought about by Moore's Law, and the speed and amount of data available via a regular Internet connection growing at much the same rate, it was only a matter of time before Algorithmic Trading would be attempted by amateurs. The marketplace in foreign currencies, for example, is full of speculators using a software platform called MetaTrader. MetaTrader allows its users to run

"Expert Advisor" software on top of the platform itself. These Expert Advisors are simple programs which monitor the market, make decisions, and execute trades automatically: just the same as the professional-level algorithms. There are thousands of small investors using MetaTrader in this way, and a community of programmers dedicated to writing new Expert Advisors. You can buy these online, from their mostly Russian or Eastern European authors. NoNameBot, for example, is an Expert Advisor that automatically trades between the Swiss Franc and the euro, and is available for $99, with a money-back guarantee from its author, "The Mother of NoNameBot, Juliya Ivanov." Not interested in NoNameBot? Perhaps the winners of the Automated Trading Championship would interest you. It's held every year, for $80,000 in prize money.

MetaTrader users working from home are never going to beat the professional algorithms at work in data-centers placed geographically just so—they're just not fast enough—but their evolution will allow them to beat amateur investors working by hand at even slower speeds. And while the geographical advantage will eventually peter out, there will always be a prize for being cleverer than the rest. This might be done by using smarter programmers, or it might be done by evolving a new and better (or even beta) strategy, as we'll explore later in the book.

Whatever the method of evolving new trading algorithms, there is a problem here. Either we're trading algorithms that no one really understands or we have a landscape of Algos working together and interacting in a way that becomes too complex for anyone to grasp.

On 6 May 2010, for example, at 2:42 p.m., the Dow Jones Industrial Index lost 600 points in three minutes. The "Flash Crash," as it became known, seems to have been the result of algorithms reacting to each other, then in turn reacting to each other's reactions, after a single large trade by an algorithm working for a mutual fund sent the market spiralling downwards. For those three minutes, to anthropomorphize, the Algos were panicking, increasingly eager to sell their holdings as the prices fell, which in turn accelerated the price drop. It was only when the Chicago Exchange's emergency system automatically paused trading for five seconds that the algorithms were able to recover their composure and stop pushing the prices down. Nevertheless, by 2:45 p.m. the market had dropped nearly 9%, with little, if any, involvement from culpable human traders. In fact, it's not surprising that no human was involved. No one could be that clever that fast. Indeed, it took five months for the Securities and Exchange Commission to investigate those five seconds of trading. Their delay was attributed to the vast amounts of data they needed to sift through. The market is too complex for anything but Algos to handle, and the Algos themselves are too complicated, at least when working together, for humans to understand.

24

REAL-TIME MAPPING

The speed of data, and our reliance on the computer systems that produce it, is not just a matter of concern for Wall Street. It's also a matter of war.

In November 2010, Nicaraguan troops crossed the San Juan river, stood on Calero Island, took down the Costa Rican flag they found there and replaced it with their own. As far as they were concerned, they had every right to do so: Calero Island, according to their map, was in Nicaragua. In their version of events, an invasion by the Costa Ricans had been repelled by brave Nicaraguan soldiers. The ensuing crisis prompted intervention from the UN Security Council, and the Organization of American States got involved. It escalated to a level described by Costa Rican President Laura Chinchilla as one of "great national urgency." The Costa Ricans thought they themselves had been invaded. The Nicaraguan Vice President denied this. "We cannot invade our own territory," he said.

The problem was, it wasn't their territory. Nicaragua had

mistaken Calero Island for its own, and reclaimed it by force. The troops had been going by the online Google Maps of the area, and the national border on those maps was, Google later admitted, 2.7 kilometers out. They *had* actually invaded Costa Rica.

In the same week as the controversy over Nicaragua broke, Google was found responsible for another dispute, this time between Spain and Morocco.

On 10 November 2010, Google Maps wrongly identified the Isla de Perejil (Parsley Island) as belonging to Morocco despite Spain having claimed sovereignty of the island in 2002, when Moroccan troops were forcibly removed. The issue threatened to ruin Spanish–Moroccan relations, until the USA was brought in to broker a deal between the two nations.

The island, less than a mile across and covered in its eponymous herb, is now marked as "disputed territory," just as the Thai–Cambodia border and the area around Kashmir are marked as disputed on atlases and maps today.

These atlases have, of course, been mostly printed on paper. That limits the extent to which they can ever be entirely accurate. Online maps are different. They can be updated continuously with new information. The fracas over Calero Island, although upsetting for all involved, could be fixed quite easily—at least as far as the map was concerned. But that's not the real value here. If maps are simply data, and that data can be added to or changed at will, then maps suddenly become a lot more useful.

They can also be made in more interesting ways. Take driving. Drivers can create maps for each other as they go. For example, the Waze system uses a smartphone application to gather information from its users in real time and distribute it to everyone else. If you're navigating through a town with a Waze app on your phone, and you hit a traffic jam, then the app will automatically send the news of your delay to all the other local users. You can add speed cameras to the map, and local points of interest too, and in countries where Waze has no mapping data to work from, Waze users help build up the map itself: by definition, wherever you drive while using Waze is a road, so the system can start to build up a framework of the new city you're travelling through. With enough journeys from enough users, Waze doesn't need a cartographer at all. The map simply emerges, and continues to change as the roads themselves do.

Getting real-time data on traffic jams in the area you wish to drive through is a fantastic service. It might actively change where and how we travel. For cities without a car culture, real-time mapping can be even more effective: public transport systems can be fitted by the local authorities with systems that report their exact position to a centralised system. Knowing the precise location of the bus you want to catch means you know exactly when to leave your front door, which is always convenient, and very necessary when it's extremely cold—systems to alert students of the imminent arrival of the school bus are very popular in the dead of a Scandinavian winter.

Real-time mapping allows citizens to make decisions about their journeys in ways that are entirely responsive to the state of the city at that very moment. While national borders seldom change, and so a mistake on a map is likely to start a skirmish, cities are always changing, and the way we react to them depends on our ability to understand their condition. While long-term residents used to have to learn which times of day, for example, to avoid the metro, today we are beginning to be able to feel the pulse of the city via our smartphones. The addition of sensors around cities, and the feeding of the data they collect back out to applications on the Internet, is what makes up the core of the "Smart Cities" idea that many technology companies are keen on. Sensors measuring temperature, noise level, pedestrian density per square metre of pavement, air quality or light levels could—if fed in real time to a map—allow people to optimize their lives as they go, as smartphones get smarter, and data connections faster.

But while representing the past and making decisions about the present is useful, Real-Time Mapping can also make predictions about the future. Google have found that by mapping the geographical origin of searches for words related to having the flu—"cold remedy," "flu indications"—they can see a nearly perfect correlation with the official flu epidemic figures collected by doctors in those same regions. Those figures, though, are collected after the fact, while the Google figures arrive in real time. Health agencies may soon have indications of groups of incidences of diseases as they happen, simply by mapping where people are searching for the symptoms online—all the better for containing and treating the outbreak.

25

SPIMES

Here we're heading into the realm of the currently theoretical but, thanks to Moore's Law, soon to be actual. A Spime is an object that can interact with the world by tracking its own process of production and by gathering information about its usage. The term was invented by Bruce Sterling, pioneer of cyberpunk science fiction, at a 2004 Los Angeles conference on computer graphics. Essentially, it's a self-documenting object.

In order to understand what a Spime might be, it helps to discuss familiar objects that have Spime-like qualities. An old book, for example, contains accumulated layers of information about the way in which it has been used: the pages fall open to your favorite passages; the spine is full of sand from your beach holiday last summer; the smell of the paper reminds you of the time in your life you bought it.

The value of Spime-like objects is increased by the additional information they embody. This is true on a sentimental level, with your favorite book, but also applies to, say, the history of their production. A canonical Spime would be identifiable and trackable, perhaps through being fitted with a radio-frequency identification device (RFID), but even today a bottle of fine wine can be Spime-like thanks to the information on its label. The value for the consumer is dependent on knowing the name and location of the terroir, the year of production, the specifics of the grape variety or blend. A huge part of the value of the wine comes not from its taste or physical attributes, but from knowing its provenance in detail. The value is in the information just as much as in the plonk itself.

Spimes are the complete opposite of the mass-produced object, identical and anonymous. This has obvious appeal to consumers of luxury goods. Let's take another example. Imagine a product made from Japanese selvage denim. The value of this denim increases with its wear. Its folds, creases and color variations determine its price and a pair of jeans with three years' worth of usage is far more valuable than one with only three months'. So we introduce a gimmick: a customer in London or New York might be able to purchase a pair of jeans and then have them worn in by Welsh hill farmers, who accomplish the time-consuming process of acquiring the desired patina of age on your behalf, safely unobserved by anyone except you, the buyer. Perhaps you could track your jeans' literal journey using a built-in GPS unit. And while you're browsing your jeans' personal profile, you can also explore all the other stages of production, from the precise chemical formula of the dyes used in the manufacture of the fabric to the name of the guy who picked the cotton plants. By the time you take delivery of your jeans, you will know everything about how they came to exist.

This might sound merely whimsical or even a little decadent, until you consider the mantra of Consume Less, Consume Better. Buying things whose value is primarily composed of information or experience rather than raw materials, has an implicitly sustainable agenda. Rather than buying mountains of disposable cotton clothing—which may well have been associated with child labour, environmental pollution, carbon release etc.—buying one pair of jeans whose value increases as you wear them, year after year, starts to seem pretty sensible. Besides, it'd be a boon for all those fashion-forward Welsh farmers.

In any case, a Spime is not necessarily a luxury product, and the implications of being able to track the means of production of an individual object go beyond the realms of satisfying denim fetishists, however earnestly neo-minimalist. A Spime could be something as humble as a banana in a supermarket. As long as an object is trackable, it can gather data, and of course, all products sold in supermarkets are trackable via their barcodes—the entire product cycle that underpins supermarkets' business models demands it.

But what if you, the consumer, could also access that information? With a quick scan via an app on your smartphone you could determine whether the plantation that grew the bananas was unionized or whether the producer had a policy to reduce their impact on the environment. If you didn't like what you found, you could choose a different banana. Such product comparison apps already exist, but are solely geared to establishing the cheapest price point. It won't be long before someone, somewhere, is imposing Spime-like qualities on supermarket bananas by layering more information to produce a comprehensive picture of their production. This would put power directly in the hands of the consumer, who would no longer be dependent on corporations' PR departments and their greenwash.

Aside from the implications of shrinking the distance between producer and consumer, there are other potential benefits to Spimes. Any object that records information about its own manufacture is going to be incredibly useful in terms of iterative design. Flaws and faults could be pinpointed with complete accuracy, the information fed back into the design process seamlessly. Safety records could be improved; the quantities of energy or materials required could be reduced. And if an object knows how it came to be, it also knows how it should be broken down into components for reuse and recycling at the end of its life.

A Spime inspires its producers, handlers and owners to more efficient, ethical and all-round exemplary behavior by recording every interaction it has with the world. But as we'll see in the next chapter, it's not only consumer goods that leave a trackable trail in their wake . . .

26

DATA SHADOWS

If you had been born a hundred years ago, unless you were rich, famous or notorious, your life would have passed relatively unrecorded by either the state or the media. There would have been a birth and a death certificate, and perhaps a marriage one. Some school examination certificates, employers' records of wages paid, maybe an article in the local paper about your prize-winning marrows or your speeding fines. There might be some wedding photos and a handful of snapshots from seaside holidays. These records were printed on paper. They got lost or thrown out or became illegible as the ink faded.

A similarly ordinary person going about their life today would generate more data in one afternoon's trip to the supermarket than our friend from 1912 could produce in a lifetime. By the time your image has been captured on a dozen CCTV cameras, you've used your debit card at a cash point, called your wife from your mobile and swiped your Clubcard at Tesco, you've generated a swarm of data variously accessible to the State, innumerable corporations and the media; one that can be stored indefinitely. That's your data shadow, and these days we've all got one.

It's not difficult to understand why many people's immediate reaction to this is negative. Hundreds of anxious or angry articles have been written about the "surveillance society" we inhabit in Europe and the United States. The average Londoner is allegedly caught on CCTV camera somewhere between 70 and 300 times a day, depending on who you believe. The fact is, though, that in a liberal democracy the vast majority of the population remain, for all practical purposes, utterly anonymous. It takes hundreds of experts weeks or even months to identify individuals from CCTV images, even when they are looking hard for something and someone very specific, as the investigations after the London bombings of 2007 proved.

We'll come back to the worrying implications of the ever-growing data shadow (especially for those people who do not live under benign governments) in due course. But it's worth remembering that your data shadow can also be a good thing. It's extremely useful when Tesco uses that data to personalise its service to you. Recommendation lists on everything from Amazon to Facebook can point you in the direction of books, music and services that you might otherwise never have discovered. Or, on a grander scale, if your NHS health records are shared with pharmaceutical companies, the medical research that is facilitated by this information-sharing might one day benefit you directly—and in the meantime it's contributing to the aggregate good.

Neither is it simply the case that you passively generate data that flows one way: into the hands of corporations or the government. Individuals are increasingly able to manipulate their own data shadow and to overlay it with previously inaccessible information held by the State and commercial entities. We'll discuss the Quantified Self later in this book.

Despite the benefits produced by our data shadows, there's no doubt that we worry about them. Often this boils down to our fear that we are no longer in control of private information. If our health records are available online for use in medical research, how long before the insurance industry gets hold of them? If Tesco knows exactly how many bottles of wine and packets of cigarettes we buy a week, how long before the NHS has access to that information, checks it against the little white lies we tell our GP and sends round the district nurse for a chat about our lifestyle choices?

Privacy is a hotly contested notion in the twenty-first century. We might feel that it is under attack, but we also have to take some responsibility for eroding it ourselves. As well as the kind of information that may be acquired without our knowledge, there is also the data that we self-publish. Thanks to social-networking sites such as Facebook, we can indulge in our impulse to record our own lives twenty-four hours a day. This generates a data shadow of things that might be better forgotten: drunken student antics photographed with smartphones and posted for the whole world to see; rash comments on a friend's status update made in the heat of the moment and later regretted. (And that's before we consider the way that social networking has made it easier to conduct, and correspondingly to uncover, inappropriate flirtations or affairs.) To what extent should individuals be allowed to edit their own past? Is there such a thing as the right to privacy of one's future self? Graduating students are now routinely advised to vet their Facebook profiles for incriminating evidence before they attend that crucial job interview.

Online privacy has become an issue requiring not only individual vigilance but also official legislation. The European Commission has passed a directive that will come into force in 2012 that enshrines the Right to be Forgotten. You will be able to demand that any site that holds data about you should delete it. But even if you delete something, it is still likely to be logged somewhere in cyberspace. Everything is traceable eventually. And in any case, isn't it antisocial to protect one's own privacy when that means reducing the data pool available for projects such as medical research, or conflicts with the public's legitimate right to know about one's actions? These distinctions are very much still being considered by society at large.

For most of us, these questions are never that pressing; our lives are simply not of interest to anyone beyond our immediate social circle. If you live in the public eye, though, your ability to control access to your data shadow is now virtually zero. A prospective Mayor of London can be certain that any over-zealous remark made as a student politician thirty years ago, *will* come out, and might destroy them. Arguably, that's a good thing: if someone's a crypto-fascist, there's no way they can keep it hidden. Then again, it might be a bad thing: most people are idiots when they're nineteen years old, but do they deserve to be held accountable for the rest of their lives to their nineteen-year-old self? Do we risk losing a potentially great public servant because of something their foolish former self once said or did?

We who live in counties where the rule of law is respected still need to monitor the use of the data shadow: after all, one never knows how a political situation will change. But it is those who live in countries with faltering democracies, or a totalitarian state, that have most right to be concerned. It isn't difficult to imagine abuse of the data shadow on a catastrophic scale. It is a commonplace observation that the Holocaust was dependent on efficient data-gathering. These days, any major supermarket could compile an accurate census of a target group, with names and addresses, in a matter of hours.

However, invoking the Holocaust is a notoriously cheap way of scoring points (see our earlier discussions of the Online Disinhibition Effect). What we're really talking about here is mass data capture, and indeed, nightmare scenarios aside, the lived reality is currently relatively banal for most. But the data shadow invokes all sorts of questions about the rights of the individual versus the aggregate, and the ethics of the social contract in a time of ever-growing access to previously private information. It's a grey area, and it's likely to get murkier, given that our collective understanding of agreed online etiquette and our legislation lag far behind the technological advances driving the debate.

27

THE IMPOSSIBILITY OF FORGETTING

As we've seen, in the twenty-first century we are all subject to the kind of documentation of our every move or thought that previously only people like career politicians or celebrities had to contend with. The difference is that, to a large extent, we are now the authors of our own record. Never before have so many moments been captured and shared by so many. We've already looked at some of the consequences of this spiralling data shadow. Its existence raises serious problems in both public and personal lives. Let's now look at these in a little more detail.

The Republican nominee in the 2012 election campaign can expect to have every single scrap of his record examined. This has always been the case. But now, because the technology exists to make it easy, that record will extend beyond policies enacted as governor, beyond the statements of his political career, to his undergraduate essays and beyond. We have a political culture that has hardened into hatred of anything that could be labelled "flip-flopping." All statements and actions are taken to be equally representative of an essentially unchanging stance. Consistency is valued above all else. The pernicious fallout from that particular expectation is not hard to identify: if you are never allowed to change your opinion, even in the light of new information, you are never going to develop or finesse your world view, which some might argue was crucial to the business of running a country in a complex world.

It's not just politicians who have to worry. They are at least already used to living scrutinized lives. The ease with which anyone can now search through that vast data shadow means that for all of us, no chance remark or youthful experiment is ever forgotten. This pressure on an individual emerge into adulthood fully formed (or else pay the price) has not yet been modified by any concessions to the impact of new technologies.

Social relationships are predicated on the ability to selectively ignore previous behaviors or choices, our own and others', that don't fit with the needs of the present. Dating in the twenty-first century, for example, is made considerably more complicated by the fact that when you arrive to meet someone, you can almost guarantee that they have Googled you already. After all, you've Googled them. So now you know they wrote for the *Socialist Worker* newspaper at university. Or stood as Tory candidate in the student elections. Information that previously they would have elected to release or withhold until such time as they saw fit is already in your possession, and already affecting your decision about whether or not to go on a second date.

Some things appear online without our consent: the back catalogue of a student newspaper, for example. Other things, like photos of drunken parties or snarky tweets, are self-published, and therefore somewhat different. But the extent to which all virtual space is now public space is not yet widely understood by anyone over the age of twenty-three.

The impossibility of forgetting in a networked world has grave consequences for our ability to reinvent ourselves at moments of emotional crisis, or new phases of life. And as well as the necessity to forget things from the past, there is the equally vital necessity to maintain distinctions between different social groups in the present. In the old days this was understood: you might not want your work colleagues to see photos of you on your stag do with your Rugby mates. There was a social understanding that that was reasonable. Now, that social contract has been smashed to smithereens by social networks. Facebook is the equivalent of inviting your Rugby mates to pop into your company headquarters, stand on chairs in the middle of the office and recount, at length, the anecdote about the jester's costume, the sixth round of tequila slammers and the waitress at the casino. With visual aids. And the things is, you have colluded in this insanity.

We know that the impossibility of forgetting is already causing problems. HR departments routinely check job candidates' online profiles. The vertiginous compulsion to cyber-stalk one's ex can make heartbreak even more painful. It is all too easy to become reliant on googling everyone, from exes to potential partners, colleagues to the new next-door neighbor. Never before have so many clues been so available. Of course we want to follow them up. Our curiosity overcomes our suspicion that it might be better not to know—for us, as much as for the other person.

Human beings have evolved to form a few long-lasting bonds and many brief, flexible ones. We have no idea of the impact on our mental health of being unable to forget even a transitory relationship, or of knowing too much too soon about our new lover's past. But as the first generation of politicians to have grown up with social networks enters office, or the first wave of kids who grew up on Facebook settles down and forms long-term relationships, society will have to evolve a code of best practice for dealing with the fallout of the impossibility of forgetting.

28

REBIRTH OF DISTANCE

Back in the early days of the Internet, one of the first things that got commentators excited, and occasionally anxious, was the idea of the "death of distance." Given that all websites are the same distance away from their users—i.e. only as far as your computer or, these days, your smartphone—physical location ceases to be the defining factor for what is available or possible. For example, pre-Internet, we could only choose from the newspapers in our local news agent. Growing up in rural Leicestershire, I couldn't access the *New York Times*, or read the *South China Morning Post.* Now we can read papers from all over the world online, and the only things limiting our access are language, or a pay wall. The *NYT* is perhaps even slightly closer than, say, the *Guardian*, if only because it takes fewer keystrokes to enter the web address.

This has significant implications. Greater choice for the reader means greater competition for the newspaper industry, of course: rivals are no longer just those other titles on the news agent's shelf; they're every English-language publication in the world. This is one of the many reasons why newsprint media are struggling to adapt to the current digital reality.

Alongside the impact on physical distance came the shrinking of time. In an age of email, communication across vast distances is essentially instantaneous. It is also free. It used to be the case that it cost more to make a long-distance phone call than a local one. By 2003, phone calls made over the Internet—via Skype, for example—were free, or very cheap, to anywhere in the world. Relationships could be maintained constantly and at no cost. This accelerating process opened up exciting possibilities: you need never speak to your boss in person; you can watch your nephew take his first steps, even though he's on the other side of the world. Many young people are growing up with no concept of a long-distance call ever costing more.

Add in the fact that as the Internet advanced, people began to think of cyberspace itself as a physical location and it's no wonder that, as far back as 1998, Kevin Kelly declared in his book *New Rules for a New Economy* that though people might inhabit places, the economy would increasingly inhabit a space.

It's certainly true that the consequences of this collapse in a conventional geography of space, time and cost have been profound. But the idea that "geography is dead" is an old one that's been around at least since the invention of the telegraph, and recently it has begun to look as if particular locations in the actual world still matter very much to the way people live their lives.

This shift is really a change in the way we fantasise about the effects of the new psycho-geography. At the dawn of the Internet, the collective dream was articulated by its early adopters, and those people tended to be highly educated middle-class baby boomers. For them, one of the most obvious and exciting benefits of the death of geography was the opportunity to leave the city and the rat race behind. Their assumption was that if people didn't need to come together, they would choose not to. They would work from home and transfer their social lives to cyberspace. This was, again, a new spin on an old theme. Home-working has been one of the ideals of futurists ever since the end of World War II and this late 1990s' version was typical of the way that technological advances can generate weirdly nostalgic aspirations. It was an essentially suburban idyll of the good life, in a cottage with roses round the door, and without the horrors of commuting to the grim city—the digital equivalent of the Metro-land developments of the London Underground. It even had a name, unusable today I feel: "Telecottaging."

There's no doubt that the freedom, for some of us, to choose where to live without the need to take into account the location of our job, is one of the huge benefits of our new "geography-lite" world. But there's an unsettling element to this dream. If people can be economically productive from their homes, that might be great for individuals who have the skills to choose to work in that way. But encouraging people to remain in their own separate units is also fundamentally very socially conservative. It reduces the creative and radical effects that develop from people mixing, as they do when they meet each other in cities. At some point, the revolution that was the Arab Spring had to step out of cyberspace and onto the street in order to be effective. And even if it's fun on your mind, rather than regime change, sometimes you still want to do it in person, with other real people, rather than via YouTube.

This 1990s' fad for trumpeting the irrelevance to people's lives of specific real places now seems old-fashioned. Yes, we are still excited by the idea of not having to trudge to the office every day, but it turns out that even if we can work at home, we still want to pop out for coffee and a chat with like-minded folk when we need a break, or feel a new idea coming on. There has been a rediscovery of the power of the neighborhood to draw certain types of people, those who are the drivers of activity in post-industrial economies. These are the people urban theorist Richard Florida dubs the Creative Classes: not just engineers, artists, designers and writers, but anyone whose job is dependent on coming up with creative solutions to problems. As we saw in the chapter on Spatial Fix, they tend to cluster together in technologically enabled areas and to be attracted to tolerant, open communities of talented people. In other words, they want to live and work in a city, and more specifically, in certain creative quarters within cities.

For these people, the Internet augments real space, it doesn't replace it. The idea of the death of distance looks to the inhabitants of Williamsburg in New York or Shoreditch in London like a retreat from everything that is most vibrant in life. Far from a hollowing out of the cities as we all flee to the cottage or the beach, the opposite is occurring. Bohemian cities fuel creativity and attract people from all over the world—a process that's accelerating all the time. Now, with location-aware mobile devices, and with your peers just round the corner in one of the creative clusters, you can all be working on the next stage of the project or having lunch, or both, within minutes.

Interaction in real space and time is far more important to commercial and social relationships than the proclaimers of the death of geography had bargained for. Human beings like to talk in person, and derive a lot of benefits from doing so. Skype conferencing from the beach is all very well from time to time, but the cliché of the hipsters holding their meeting in a café with their MacBooks is a cliché for a reason. Chat is a much higher bandwidth activity than we realized Real places, at least of some specific types of real places, are here to stay.

29

LIVE PERSONAL-BEHAVIOR SHARING

One of the recurring anxieties attached to the shift to digital was that it would be dangerously isolating. If social interaction and work could be carried out via your screen from your own home, why would you ever get out and speak in person to anyone ever again? A de facto privileging of the textual content of messages, divorced as they were from the contextual nuances of tone of voice or body language, would also coarsen communication. This anxiety has turned out to be largely exaggerated. Yes, many children spend far too long in front of screens and may in consequence struggle to relate to others in the flesh, but that's a parenting issue, not a technology one. And in any case, as they grow up they're likely to discover that there are other rewarding things to do that require solid-world interaction. We are all optimising our Internet usage all the time, and the vast majority of today's heavy users will find a way to moderate their intake in due course. That's the subject of another chapter; for now I'm interested in the way in which the Internet is itself evolving its own ambient communication. The principle mechanism by which this happens is live personal-behavior sharing.

There's a growing class of apps that are used most often via smartphones, which continually monitor and make public certain small aspects of users' behaviors. They are location-sharing apps like Foursquare, or activity-sharing apps on Last.fm or Spotify that let your contacts know what music you're listening to at any given moment. Even a status update on Facebook or Twitter is derived from the same notion of live broadcast, though with a higher degree of direct input from the user.

The immediate impact of such apps is to foster first-degree social bonding. If you see that a friend has checked in via Foursquare at a café round the corner from your house, you could pop in and join him for a coffee. If you notice that your brother, whose musical taste you generally share, is listening to a band you've never heard of, you could decide to check them out.

The secondary impact, and arguably the more interesting one, is the comforting sense of connectedness that derives simply from knowing what someone you care about is up to. You can share in their everyday life as they move around the city, for example, receiving bleep alerts as they go. It adds up to a curiously intimate sensation of being connected, but with the lightest of light touches. One of my friend's Foursquare check-in as she arrives at Finsbury Park station now functions as my daily prompt to start my day. I enjoy the sensation of knowing that she's getting on with life. If that morning bleep were missing a couple of days in a row, I would send a text to check she was okay.

By using these apps we gain extra senses that are hard to quantify but can be powerful. Changes to our friends' routines, to their arrival time at the station, or to the frequency of their tweets, can alert us to changes in their mood faster than a verbal message might. Digital communication is developing its own equivalent of the tone and body language clues that some people worried were irreplaceable.

This online ambient communication is actually more consistent and more far-reaching than the flesh-world version, because it's always with you, wherever you are. A nod in the street to an acquaintance, a brief chat with your co-worker as you pass them in the corridor, these are forms of social glue for sure, but then so is watching the world waking up every day as your friends in San Francisco and Tokyo come online, or registering that your aunt in Denver must be ordering breakfast in her favorite diner as usual.

As it becomes more and more usual for families to scatter around the world, these low-level check-ins are another way to stay in touch. They might not be as crucial as free video calls via Skype, but the way in which they mimic the background awareness we accumulate when living close to someone is subtly powerful. It's more like social weather than anything, locating us in a global flow and allowing us to connect intuitively with the people we love without breaking off from what we're doing until we're ready to.

Another great spur to live personal-behavior sharing is the phenomenon of the Quantified Self. As we'll see, one of its key aspects is the motivation that results from, say, having a feed to your contacts' phones that registers your hours in the gym or calorie intake. If you are in a group of dieters who share automatic updates in this way, you can encourage, support or chide each other as necessary.

The potential of this ambient communication is only now really becoming apparent. Its next evolutionary stage might focus on broadcasting social cues that enable other people to relate to you in the way you would prefer. For example, if you have a deadline for a huge piece of work coming up, it might be useful to have a subtler tool than an out-of-office message to tell your colleagues and contacts that unless their enquiry is genuinely crucial, you won't be getting back to them until Monday. For the moment "busy" icons (on Google Chat or Yahoo Messenger, say) will have to suffice.

The Internet has been one giant experiment right from its inception, and now that there are billions of users worldwide, any app is constantly evolving to meet its users' needs. It's only through this usage that we can determine what something is really for, as opposed to what its designers thought it was for. Many apps don't survive, but the ones that morph from their original proposal to the version that fits best with what people actually want are small things of beauty. It's always been a lovely thing to feel your friends' presence around you, now you can even if they're on the other side of the world.

30

THE QUANTIFIED SELF

Eagle-eyed readers will no doubt have spotted two of the core ideas that I keep returning to: the impact of iterative design and the tendency of our digitised lives to generate vast amounts of data. Many web-based businesses are situated at the precise intersection of these two ideas. Facebook, for example, employs iterative design to mutate constantly, in response to the huge amount of data it collects from its millions of users. Amazon does too.

One area of human endeavour that has so far remained relatively untouched by this intersection of concepts is medical research. That's because such work takes a long time to become statistically viable. You can deduce that smoking is pretty bad for us by asking a non-smoker to work their way through a packet of B&H in an afternoon; but if you want to prove that smoking kills a good proportion of smokers, you need to track thousands of subjects for several decades. Even then you will be dealing in averages and generalities. There is, however, a micro level at which new technologies have enabled people to draw very precise conclusions about their own wellbeing via a sort of DIY medical research.

Back in about 2008 a group of people (for which read: technology alpha geeks) realized that by using smartphones to take measurements of their mental and physical state throughout the day (and night), they could begin to analyze, and optimize, their health. The impact on weight, fatigue or mood of various factors, such as the hour they'd got up that morning or what they'd eaten for lunch, would be trackable in ways that were almost automatic and certainly highly compelling. Social networks could then be used to share the data with other participants in the experiment.

Before long it became clear that a small number of users following basic empirical principles could arrive at conclusions that were both generally consistent and still very personal. And they could do so within a matter of weeks and months rather than years. For example, by tweaking the calorie content of their lunch menu every day for a fortnight, individuals found they could raise their energy levels and sustain their weight loss. What's happening here is that small changes are being made often, in response to (relatively) large amounts of data. In other words, the notion of a Quantified Self is to personal wellbeing what iterative design is to web-based businesses.

Unsurprisingly, a whole industry of tools and apps emerged to market the practice to a wider public. Now there's a raft of products, each one basically a combination of pedometer to measure movement and an app or website to enter additional data so that you can also count calories or record your mood.

Personal psychology is of particular interest; partly because it's one area of wellbeing that is especially hard to assess by relying on memory alone. The types of conditions that afflict many people in the developed world, such as stress and anxiety, are also very sensitive to identifiable triggers. While traditional medical research is obviously useful background reading, it can't illuminate your own personal triggers in the same way that, say, a heart-rate monitor strapped to your wrist, or a diary of meals eaten, can.

This process of gathering, analyzing and sharing your data can become highly addictive, particularly to the game-savvy techie types who are its core proponents. As well as potential improvements to their own health, there is also the thrill of competition (a website that broadcasts your weight every time you step on the scales, anyone?) and the chance to contribute to the aggregate good. If I eat lots of sweets in the morning, I'll notice that I'm rather grumpy in the afternoon—I'll only notice this if I write down both my sweet intake, and my mood—I can then both personally cut down on the Haribo, and add to the more universal set of knowledge linking blood sugar spikes to mood swings.

The Quantified Self (QS) movement believes that it is crowdsourcing medicine, via online communities, meet ups and conferences. Its practitioners are highly motivated and committed to harnessing precise, accurate data about an ever-expanding number of conditions, from asthma to depression. Naturally, this is the kind of data set that various entities in the pharmacology, diet and wellbeing industries would love to get their hands on. QS groups are attracting more and more sponsorship and the technologies are being incorporated into more mainstream product ranges such as Nike+. But the hardcore aficionados have already moved on from pedometers and calorie counters and are pushing the limits of what can be measured and how accurately (blood sugar and lung capacity are up next).

Quantified Self is a nexus of many of the ideas featured in this book. It links data sharing, iterative design and, as we shall see, gamification at a personal body level. The result taps into the most fundamental principle of the new technologies: a flow of power away from the old elites and centers of expertise. We have grown accustomed to citizen journalists over the last few years, but with Quantified Self we are now seeing the rise of citizen scientists.

31

PERSONAL GENETIC TESTING

The powers of the citizen scientist don't stop at simple data collection—they now run much deeper. In fact, the technology for mapping your individual genome, at least to a level of medical interest, is now so cheap that it's available online for $99. The science that was unimaginably groundbreaking in 2000 when it was announced that the human genome had been mapped for the first time, the test that cost $1,000 just five years ago, can now be purchased during your lunch break. You send away a sample of saliva and the scientists work out your statistical risk of Alzheimer's or skin cancer. That's Moore's Law in action. But is it such a good idea to have an answer to the question, how likely am I to develop a serious illness? Especially if there is nothing you can do to influence your prognosis, as is the case with Alzheimer's. And what about the implications of other people getting hold of the information?

Firstly, there is an obvious limitation on the usefulness of this technology, even to the individual. Very few diseases can be predicted with any degree of statistical certainty by examining genetic vulnerability alone, and very few diseases can be linked to a single gene. A truer result would take into account lifestyle factors and would allow for the fact that combinations of up to forty different genes are thought to be indicators of, say, diabetes. There are exceptions. The presence of the BRCA gene means that you are almost certain to get a particularly aggressive form of breast cancer. It is considered to be such a definite indicator that carriers are advised to have a preventative double mastectomy, which although extreme might be a solution. But a more typical result is the one I received. I have a slightly higher than average risk of lung cancer, so I should probably lay off the cigars. Then again, I already knew that.

If you are in possession of the information, you are at least able to make your own choices about how to respond. But some health professionals don't believe that commercial genetic testing is good for people at all. They argue that, aside from confirming the standard healthy living advice we already know, most of the information delivered by these tests is the kind of news that people should hear from a doctor, with access to specialist counselling. Otherwise they run the risk of severe psychological damage.

Moving on from an individual's right to know, there's the tricky question of how to control access to the data by other potentially interested parties. In the future, perhaps some people will demand pre-nuptial genetic testing. Maybe we will routinely undertake genetic testing before we have a baby with our partner. What choices might we have to live with if the results came back unsatisfactory?

The information about our genetic fallibility is clearly of enormous interest to insurance companies and banks, and indeed to anyone with a vested interest in when we might die. It's not difficult to imagine that a bank might withdraw its offer of a mortgage from anyone who refused to take a genetic test. Similarly, employers might make their job offer dependent on your results. The possibilities for genetic discrimination grow with every new condition that is attached to its genetic marker. A propensity for bipolar disorder and schizophrenia can now be tested for. It's possible to imagine a scenario in which an individual's life could be severely curtailed not because they actually *were* ill, but because of a theoretical increased propensity to, say, depression.

So far, the consensus has been that nobody wants that to happen. The concern is such that in many states across the U.S., direct-to-consumer genetic testing is heavily restricted. Genetic discrimination is illegal. The data is ring-fenced and insurance companies are banned from using personal genetic testing to determine the cost of their policies. But as with many of the issues thrown up by *Approaching the Future*, our ability to develop protective legislation and new cultural norms of behavior is much slower than the rate at which the technology is advancing and the data is becoming available.

32

BIOHACKING

Genetic engineering is one of the most exciting scientific breakthroughs of recent times, one that inspires all sorts of conflicting sentiments: hope for salvation from diseases; anxiety about the risks of cloning. Thanks to the huge amount of coverage it generates in the press we have a sense of what it involves, but the label itself is misleading. The experimental processes carried out by geneticists are actually rather more like cooking than engineering.

The explanation for this lies in a sadly underappreciated aspect of the success of the Industrial Revolution: standardization of parts. Previously, if you wanted to make anything mechanical, each tiny piece had to be handmade. This was fine until a component broke, or you wanted to tweak the design of the machine. Then the replacement part had to be made precisely the same as its original, which was fiddly, time-consuming and expensive work. Under pressure to produce the innovative ideas that were emerging, manufacturers devised standard sizes of nuts and bolts and from there, endless other components. The impact on mechanization was dramatic; a creative revolution was unleashed and the applied science of engineering was given the tools it needed to become the driving force of its age.

What we understand as genetic engineering involves the manipulation of an organism's genome to produce a desired effect. Typically this is achieved by introducing DNA snippets from a different organism, or by removing a particular DNA sequence. The thing is that the desired effect can be hard to achieve, at least on a reliable basis, precisely because a DNA sequence is not analogous to a standardised part. Relatively speaking, genetic engineering is tinkering. Or at least, it was.

Tom Knight is a senior research scientist at MIT, where he has been building pioneering computer hardware since the late 1960s. He became fascinated by biochemistry and genetics in the early 1990s and has since been on a quest to develop what he calls BioBrick parts, i.e. standardised components. A BioBrick part is a snippet of DNA that we know for sure has a specific effect, that can be mass produced and handled in a consistent way to produce replicable results.

In 2003 Knight and his collaborators launched partsregistry.org, an open-access collection of genetic material that can be ordered by researchers anywhere in the world for delivery to their lab. Partsregistry.org is a classic Internet-based community resource, open to all and reliant on the collaborative spirit of its users to enrich itself. The hope is that BioBrick parts will transform genetic experimentation from biology to true engineering, with a consequent explosion of creative solutions to all sorts of problems.

Of course, a side effect of this high-level open-source work is that amateurs are increasingly getting involved. This science is no longer the exclusive preserve of research geneticists: it has already become part of the hacker culture. With some basic knowledge and three or four thousand pounds, you can now assemble the necessary equipment for a home lab by shopping on eBay. With the costs of DNA sequencing tumbling, the facility for ordering genetic parts online and the enormous amount of open-source information available, genetic engineering is set to go the way of car mechanics, home electronics, computer programming and digital media and transform itself into an activity for increasingly skilled and dedicated amateurs. The difference in this instance is the speed with which the tools and (at least some basic) expertise have been transferred across an unprecedentedly large skills gap, from university-level professionals to the passionate hobbyist.

Already in the States there are biohack spaces running workshops for school groups. The children might be learning how to introduce the DNA sequence that makes a glow-worm glow into an E.coli culture in order to produce glowing bacteria. Or perhaps they'd rather make bacteria that smells of lavender. There are even prizes for the best genetic hack.

One of the enduring cultural anxieties about genetic engineering is the spectre of a lunatic individual, or perhaps these days a terrorist cell, capable of creating an organism, say a bacteria strain, that would be detrimental to human safety. This is the same recurring anxiety about technology becoming available to those who "should not" be allowed it that has been haunting our imaginations since Mary Shelley published *Frankenstein*—if not before. In 1818 the Industrial Revolution had settled in to its stride and was bringing with it a complete reorganization of society. That level of change was anxiety inducing, and implied winners and losers, even without murderous automatons. Nowadays there is undeniably greater potential than ever before for people to misuse technology, but the fact that some people could use it for harmful purposes does not remove the greater good that arises from wider access to new technologies. Then there's the fact, touched on over and over again throughout this book, that shutting the new information and tools away is in no way a practical possibility.

Genetic engineering is still very much a new scientific technology with virtually untested potential. Its professional use is likely to produce extraordinary breakthroughs in the next few years. But its morph into biohacking will also throw up all sorts of new ideas. So far it's been mostly gimmicks—bacteria that smells like a florist's shop—but everything we've seen in other spheres suggests that when a connected community learning from one another, standard parts and professional tools converge, an explosion of creativity occurs. A new revolution in biotechnology must be imminent.

33

NANOTECH AND OTHER MIRACLE TECHNOLOGIES

Biohacking labs aside, the future has turned out to be rather disappointing. Where are the jetpacks and domestic robots that were dreamt of in the Atomic Age? In many ways our world looks the same as it did in the 1950s, with semi-detached houses arranged in suburbs, motorways and gas stations for our cars. Not a jetpack in sight.

The over-excited vision of the 1950s and 1960s was one of total renewal, as if the bomb had dropped but nothing bad had happened, just an aesthetic cleansing. In reality, even the techno-fantasies of civilian nuclear enthusiasts turned out to have complicated consequences. Fortunately, for most of us the future arrives not in an explosion but gradually, layering or assimilating with the past. The Atomic Age's devotees had to convert the horror of being bombed into a new reality with their determinedly optimistic vision of a future miraculously free of the debris of the past. But by the Eighties the implausibility and undesirability of a complete reinvention of the fabric of reality had become obvious. Not that the future's creeping arrival was necessarily benign, but its gradual process did seem more likely. This is the difference between the futurist visions of *Star Trek*, all new and shiny, and *Bladerunner*, in which the future is an inevitable consequence of a decaying past.

Jetpacks haven't figured in our discussion so far. Neither have robots that do the chores, despite a surge of interest in fractional AI (more of this in the following chapter). This is in part due to technical challenges, but more to the realisation that some technologies are either not useful enough to invest in, or not socially desirable enough. Ponder for three seconds the consequences of having drunken boy racers on jetpacks as well as in cars and you'll get my point.

With the radical exception of digital technologies, even the most up-to-date consumer-technology products and the most modern design is in fact very reminiscent of design classics of the 1950s and 1960s. The Internet-derived capabilities of an iPhone are completely different from anything that was available fifty years ago, but the phone looks remarkably similar to the gadgets designed by Dieter Rams for Braun in the 1960s. The future has crept up without troubling us either with killer androids or colonies on the moon.

But our imaginations are still fired by technologies on the verge of arriving, especially if they hold out a promise of salvation. Nanotechnology is one of the few remaining pieces of the old futurism that continues to excite interest. Its promise is of machines the size of molecules that could be injected into the bloodstream and sent to hunt down and destroy viruses or clear fatty deposits from arteries. There have been vast amounts of research and there is still work being done on the assembly techniques that would be required, but for now useable nanotech remains fifteen to twenty years in the future, as it has done for the last thirty years. It is now more an article of faith that medical care will at some point be revolutionised by nanotech than a serious proposition.

It's noticeable when one starts to think about the miracle technologies, the ones that will apparently save us all, that by and large they are the fantasies of people trying to preserve the status quo. This is not always the case, of course: with nanotech, few people would argue that using it to eradicate a virus from someone's body was an essentially conservative measure. On the other hand, using nanotech to clear someone's arteries so that they can continue to eat junk food might be considered a retrograde step.

Medical ethics is full of such binds. So is environmental science. A classic miracle technology is the algae that wees petrol. For years we have been on the brink of developing genetically engineered algae that is capable of absorbing atmospheric CO_2 and sunshine and excreting biofuel. The prospect seems appealing, unless one considers that enabling the continuation of a lifestyle that is unhealthy for the ecosystem is not worth preserving. The same goes for the drive to engineer pork or beef proteins in laboratory conditions. If we're running out of the capability to feed everyone animal protein, should we grow it artificially or would it be better to change our habits and make do with less? These are all example of miracle technologies that sound as if they are taking us towards a radical future but may in fact be preserving an ingrained social, political and cultural system.

As resources come under more and more pressure we will see a great deal more of these miracle technologies up for discussion. As we discuss later in the book, long-term planning—across the generations—is something that human beings have never been very good at, but we urgently need to get better. It might also help if we could invest our excitement about the future in those technologies that imagine change as something constant and natural to be revelled in, rather than something that shoves an otherwise static society in a particular direction, or fixes it in a frozen moment of development. Using this way of thinking, investing trillions in tiny robots that allow us to carry on eating pie is ultimately not that exciting. Exploring the constantly evolving capacity of the digital revolution to change reality, on the other hand . . .

34

DIPLOMACY IN THE TWENTY-FIRST CENTURY

Old-fashioned diplomacy (as distinct from the self-interested foreign policy it existed to serve) was a very polite business. For hundreds of years it involved, at its simplest, sending ambassadors to each sovereign state to discuss matters of interest with that state's representatives. This was a rarefied practice undertaken by individuals from the very highest echelons of the ruling elite. And then, of course, there were the sort of informal international relations that happen when people trade with each other, emigrate or go on holiday to each other's countries. The introduction of Indian food to Great Britain was not the result of ambassadorial decree but the immigrant's longing for home, not to mention the necessity of earning a living. But in either case, the process by which cultures or nation states could exert influence was relatively slow: weeks or months for our ambassador; months or years for the Indian restaurant owner. It is a truism that power used to be more centralised and to move in a more stately fashion than it does today, but in few areas of life is that more clearly seen than in contemporary diplomacy.

The possibilities offered by the Internet are at first glance very exciting for anyone whose job it is to represent their nation's interests. Previously, an individual was entrusted with delicate negotiations, via meetings with key people, in the attempt to, say, convince a hostile regime to reform its voting system. If the opposite number simply refused to talk, there was very little that could be done to advance the cause. Now, in theory, it's possible to circumvent the ruling classes completely and speak directly to the population via social media. That diplomat doesn't even need to travel to his posting; the Internet can be used to facilitate the kind of mass exchanges of ideas that used to have to happen in person via long years of student exchanges or trading partnerships. This accords with the familiar idea that exposure to the plurality of alternative opinion will change people's perspectives. But as we'll see in the coming chapters, while it's true that it can happen like that, it won't necessarily, and certainly not as quickly and completely as techno-diplomats might like to imagine.

For a start, even if the populace of your target country warms to your ideas, that may not amount to much if their government doesn't. There are, after all, plenty of places in the world where the regime is sufficiently totalitarian to ignore the wishes of its people, and those are precisely the sorts of places that are likely to take a dim view of you targeting their citizens directly. Then there's also the fact that a populace can be sympathetic to your "soft power" cultural outreach work and remain hostile to any more political agenda. When the traditional etiquette of diplomacy is flouted, what looks from one perspective like information provision can feel like information imperialism from another.

A related problem for our Facebook-, Twitter- and YouTube-enabled foreign policy team is that with all this reliance on websites, diplomacy starts to look distinctly like branding. The techniques available to a nation state that wants to present itself to the world as embodying a particular set of values are exactly the same as those available to any commercial brand. The point of consumption (laptop, smartphone) is the same. But the reality is that nation states are likely to suffer by comparison with the branding efforts of any big business. The marketing budget at Chanel will dwarf the amount of money available to the U.K. Foreign Office for its soft power diplomacy. And Chanel has been in the branding business for a heck of a lot longer than the offices of the central government. Arguably, choosing to operate in the same arena as a million commercial enterprises disadvantages the foreign policy department. The Internet is a very visually demanding and novelty-loving place. There is a distinct risk that Brand U.K. will end up looking rather tawdry when it has to compete for attention alongside Apple—not because of any intrinsic lack of value, but simply because of lack of expertise and budget.

This has serious consequences for your ability to, for example, switch allegiances from radical Islamism to liberal democracy in new and creative ways that don't involve ground troops. If you use the language of the market, you might very well end up cheapening your message. After all, if you post a promotional video for the U.K. that sets out to tell the world what the country stands for, you are going to be competing for hits on YouTube with the video of the sneezing baby panda. Is that really a fair fight? Suddenly the Internet isn't looking like such a shiny new toy after all.

Part of the problem lies not in the medium, or even the discourse you use, but in the content itself. Or rather, in whether you can live up to the values you are espousing on your website. First, you have to fix on some concepts to sum up your country's identity. There have, for example, been endless attempts to define Britishness in the last fifteen years (warm beer and cycling spinsters versus Cool Britannia, anyone?) all of them nebulous and vulnerable to debunking. But once you've hit on a formula you like and sent it out into the virtual world to spread its message, you then need to ensure that you are equipped to stand by it, in the new era of transparency.

As we've seen elsewhere, the digital world is very good at measuring things accurately and uncovering the sort of detail that could thirty years ago have remained buried. That means that any statement you put online had better be independently verifiable, because it *will* be scrutinized. That's why businesses have PR departments to make sure their corporate social-responsibility strategies or their environmental best-practice codes are not being overstated. Because you can be certain that everyone from suppliers to competitors to Greenpeace will be looking for holes in their story.

How much more complex is it for a nation state to live by its set of nebulous values? And how much greater are the risks of losing the battle for hearts and minds? After all, your foreign policy is, for all practical purposes, not what you say it is, but whatever comes up first on someone's web browser. As we'll see in the chapter on echo chambers, web search results are not neutral but influenced by previous searches. If you want to be a player on the world stage, you'd better be a whizz at the art of improving your Google ranking . . .

If this all makes you feel a little twitchy, that's hardly surprising. The difficulties inherent in modern statecraft are symptomatic of the erosion of the power of the nation state, which is bound to worry many of us. As power is draining away from the old hierarchical structures, including central government, it's flowing in all directions. In this case, to regional governments such as the EU, or to supranational bodies such as the IMF. It's also dispersing to city governments. The reality is that, increasingly, it makes more sense for some activities to be carried out at these levels rather than the national one. But that doesn't have to be a problem. After all, it's much easier for a city like London or an area like Cornwall to sum up the specific qualities that make it attractive as a place to live, do business, invest. Many of the issues of being accountable for your online identity in an age of transparency are therefore much less acute for London or Cornwall than for the U.K. as a whole. There's no reason to think that our chances of getting the governments and the policies we want and need are made worse by the Internet. In fact, the reverse might be true, if we all embrace the new networked reality and wake up to the fact that the power is flowing in our direction, too.

35

MULTIPLE-AXIS POLITICS

European and North American politics during most of the twentieth century was the story of Left vs. Right. The potted version of this single-axis politics presents the Left as progressives, pushing for change, reinventing society to make it fairer, more equal, more tolerant. The Left was about extending civil rights to minorities, the provision of a welfare net by the State, the trade union movement. The Right, by contrast, has tended to be characterized by conservatism, a sense that things are basically good as they are and need only a few tweaks. The Right was about individual liberty, a small state, trust in private enterprise as the driver of growth and respect for hierarchical power structures already in existence. In reality things were never so simple, of course, but nonetheless, the dominant narrative in the developed world understood politics in these basic terms.

Towards the end of the last century, these binary political positions started to meld. There was a convergence on the center, especially after the fall of the Berlin Wall, but the center was further right than ever before. The Left gave up calling for the redistribution of wealth and dropped any associations with discredited socialism. There was a consensus among politicians from either end of the single axis that the underlying structure of ultra-liberal free-market economics allied with liberal democracy was unquestionably the best option available, both as a mechanism for governing a nation state and for running the interactions between states. The result has been monotheistic commitment to the model that has dominated since the 1980s, and a lack of debate, at least within mainstream politics, about alternatives.

At precisely the same time as this deification of an idea was leading to the widespread adoption of a basically conservative (small c) agenda, we gained a mechanism for devolving power to the masses on an unprecedented scale. As we have seen over and over again, the Internet facilitates free and easy group-forming and flow of information. It means that when events such as the global financial crisis or the Arab Spring occur they are played out live on the Web, and we can all form our own opinion, independently of traditional broadcasting, if we so wish.

The effect has been explosive. Unemployed Spanish youth watched, learned from and supported the Arab Spring and staged their own protests across Spain. When it transpired that no one on the mainstream progressive wing of the single-axis political structure in either the U.K. or the United States was going to make any serious fuss about the failures of the financial industries, the Occupy movement grew up to protest the power of the 1%. That movement went global, thanks to the Internet. Since 2008 and the first crash of the global economy, we have all had ringside seats as the supposedly sleek machine that underpins our way of life judders and slows. Even as the key players in the hierarchical power structures assure us it's business as usual, or it will be any minute, we have been able to test and to participate in the opposite point of view, to an unprecedented degree. The doctrine that states that the developed world has reached a near-perfect apotheosis of a political, social and economic system, give or take a few details, and that in any case there is no alternative, has been coming under the first sustained critique since the heyday of Communism.

And it's far more complex than merely a Leftie fight back. In fact, there really hasn't been much of a Leftie fight back, not in the old single-axis sense anyway. The people who no longer feel represented by the system that we have been assured is the only possible option are forming new political movements, outside of the mainstream parties, and getting things done. They are pressure groups that represent a larger silent minority, and they are sprouting at some interesting intersections of a new multi-axis politics.

The Tea Party in the United States, for example, is socially and culturally as conservative as they come, but it is also deeply hostile to the financial services industry. That combination of values has been literally unthinkable since Reagan's presidencies. Though it has strong links to some high-profile Republican politicians, it is a networked organization with no leader or central party and regards dissent from mainstream Republicanism and support of a local agenda as its *raison d'être*.

The Occupy movement is, similarly, a loose coalition of activists, this time of a progressive stripe. Occupy functions outside any mainstream representation of its views, and has far fewer links to party politicians than does the Tea Party. Formed of loose networks of people who have in the past campaigned for nuclear disarmament, against the construction of coal-burning power plants or the Iraq invasion, Occupy's call has been for a reconnection between a Leftist social agenda and a rejection of a pure free-market economy. As with the Tea Party's paradigm-busting values, that particular combination hasn't been seen in Democrat or Labour circles since long before Bill Clinton or Tony Blair's terms. Back then the Left would have been unelectable if it hadn't pledged its faith in the ultra-liberal market capitalist dream. But since the dream turned sour, new questions can be asked.

From the opposite ends of the old spectrum, these two groups may have hit on a truth of our age. Unfettered capitalism is highly effective at pursuing its own ends. It skews debate so that discussion in the mainstream media and culture takes place on its own terms. It's addictive to participate in when you're winning, and skilled at telling you that if you're losing, it's your fault. Except now that winners are turning to losers they are less inclined to believe that capitalism's losers have no one to blame but themselves.

What happens next is anyone's guess. The break-up of the euro, the invasion of Iran: at the time of writing there are several potentially cataclysmic events on the horizon. There is certainly a hunger among many voters of various political persuasions for something that does not exist, a politics that takes into account new ways of affiliating and representing. The time is ripe for new leaders to emerge, leaders capable of harnessing the power of the Internet to connect directly with the mass of discontented and disenfranchised people, bypassing all the old single-axis apparatuses of power.

The story of twentieth-century politics in Europe has taught us to fear charismatic leaders who impose their fanatical programmes on the people, harnessing party politics for their own ends, or eschewing them completely, but for every Mussolini or Hitler there could be a Ghandi or a Nelson Mandela. Extremism is a real possibility of course, and a frightening one, but is not necessarily a given, and is less likely if the system adapts promptly to the new reality rather than brittlely insisting that all will be well. Whatever happens next, its shortcomings are too visible now, thanks to data transparency and citizen journalism, for it to sustain this position. For the first time in history, a determined individual (or an impassioned group like the Tea Party or

Occupy) doesn't need a structure such as the Army or the trade unions, or the machine of an established political party to come to power. Perhaps, all they need is charisma, some appealing ideas, a lot of money and a very clever digital campaigns manager. With all that in place they could be in touch with voters. Encourage them to form groups, take on campaigning, set up local chapters. They could build their own organization from the ground up. They might even find some unlikely allies in the heart of the most hierarchical establishments . . .

In Britain in 2010 David Cameron launched the concept of the Big Society, founded on the ideas of devolving power to communities, and passing public-service contracts to the private and voluntary sector. This sits well with classic small-state conservatism, but it's also not a million miles away from sub-contracting policing the Internet for child porn and hard drugs sales to an anonymous collective of hacktivists (more on this later in the coming chapters). There are likely to be some eye-opening reconfigurations if the shift towards a multi-axis politics continues.

History shows us that predictions about political movements' rise and fall are rarely better than sound bites, but one macro-prediction I will stand by is that the party politics that govern our nation states will never be the same again. If there is one thing we can say with any confidence, it is that when the Internet turns its eye on any industry it destroys it and remakes it in its own image. This is what has happened to vast swathes of the retail sector, and to the music, travel, newspaper, publishing and broadcast industries. Nobody should assume that politics will be exempt from this creative destruction. Increasingly, politics will happen online and on the street, as it did during the events of the Arab Spring and in Europe during the discontent of 2011. I'm not claiming the revolution is imminent, merely that the bit in the middle, where the politicians live, will be squeezed hard and emerge a different shape.

When Barack Obama was elected President of the United States in 2008, many commentators discussed his savvy grasp of the power of social media as a campaigning and fundraising tool. It was dubbed "the Facebook election," his Twitter campaign was hailed as a masterful shift towards a new form of digital politics. The campaign certainly was savvy, and the power dynamic certainly did change. Obama left everyone else in the dust for sure-footed grasp of the potential of the web as tool to win an election. Even so, I suspect that with hindsight, we may come to view the 2008 election as one of the last of the old-fashioned kind, before the Internet tore everything up and started to reinvent the politics of the nation state in its own, networked, anti-hierarchical image.

36

THE ECHO CHAMBER

Early enthusiasts, or at least the more idealistic types among us, hoped that the Internet would be one giant tool for spreading tolerance. Their argument boiled down to the idea that extremism of any kind was determined in large part by environment. If you were brought up under South African apartheid, you might have some very illiberal views on race. But had you been able to access foreign media and communicate freely with other people in places not under the same constraints, your views might have been altered. Given all the facts, people couldn't help but see the error of their ways.

To a certain extent, this belief in the power of the Internet to change people's opinions has been borne out. The Chinese state would not be so assiduous in censoring websites were it not worried, justifiably, that the population's increasing exposure to different points of view might undermine its ideological grip.

But it turns out that unless there is a thirst for alternative information—usually among those who feel that their lives are unsatisfactory and that ideas are being withheld from them—people rarely go looking for it. If you hold a strong opinion about something and vehemently believe that you are right, you are extremely unlikely to search out anything that contradicts your world view. Our opinions are almost always based not on reasoned examination of all available evidence, but on emotional reactions and ingrained beliefs, and most of us don't enjoy the sensation of being contradicted. On the contrary, it tingles quite pleasantly to be confirmed in our views.

This truth is typically amplified rather than undermined by hanging out online. In cyberspace it's easier to find other people like you than it is in the real world. You no longer need to trawl record shops for obscure fanzines or track down a man who might know a guy who can tip you off about the political meeting you're still a bit ashamed to attend. There's an Internet forum for everything and everyone, so you can find other fans of early-Eighties Scandinavian punk groups or your local branch of the Nationalist Socialist Party equally easily. Once you've made your way to the forum of your choice and discovered to your delight that there are plenty of other people who think as you do, a dose of the Online Disinhibition Effect will see you expressing all manner of views that you might keep to yourself if you were talking to your mother-in-law, or even to an acquaintance in a bar. Meanwhile the echo chamber works to bombard you with the message that this is all perfectly normal, because everyone else is doing the same thing. At least everyone you've ever met.

This is particularly true when it comes to political discussion forums, but applies to any emotive issue. You don't need to be lurking in the murky fringes of the Internet to notice it either. The reader comments below articles on the *Daily Mail* or the *Guardian*'s website are enough to illustrate the power of the echo chamber at work. Commenters reinforce each other's opinions and frequently ramp up each other's vitriol towards anyone whose views are different. Even trolls (posters whose comments imply that they are being deliberately antagonistic) simply serve to further polarize the debate. By weighing in with an entirely opposite argument, the dialectic is strengthened and nobody has to bother with the uncomfortable work of actually listening to a reasoned alternative. The middle ground vanishes, and appeals for good behavior are useless, since naturally no one is prepared to admit responsibility for skewing the debate with their crazy ideas.

Added to this user-driven echo chamber is a technological factor that accelerates the process. Our experience of the Web is increasingly personalized Some of that is controlled by us, by where we choose to go, but much of it happens via filters at levels we might not be fully aware of. At the benign end of the spectrum, having Amazon direct us towards novels we haven't read but will probably enjoy is clearly a great service. The way Amazon delivers that service is by examining what we have already clicked on, and offering up more of the same, or with slight variations, depending on how much of others' data shadow it can cross-reference with our own tracked searches. This is because computers are still nowhere near as good at making intuitive leaps as people. Netflix, the film rental site, ran a competition for programmers to improve their recommendation algorithm by 10%. The $1 million prize goes to show the value of a good recommendation system. But even the new improved algorithms can't deal with things like the anomalous fact that practically every film fan, of whatever persuasion, loves *Napoleon Dynamite*. Artificial intelligences just can't handle surprise sleeper hits.

This tendency to narrow our choices may be technology driven rather than a thought-control conspiracy, but of course a personalized service—or, as we could think of it, an echo chamber embedded in your search engine—is still potentially scary. Google, or any other search engine, only displays the results that it has decided you will be most interested in. That means that, in theory, it is perfectly possible that you could go through life without anyone challenging your opinion that the National Socialists were the future of politics. Since we don't live in cyberpods, this is currently unlikely, but we already inhabit a virtual landscape that reconfirms our opinions constantly. For example, the growing personalization of search results means that two people searching for "British Petroleum" might get very different results. If one has previously searched frequently for, say, Greenpeace, they could be presented with stories about the 2010 oil spill in the Gulf of Mexico. If the other reads the *Daily Telegraph* online every day, they'll be directed to BP's stocks and shares performance.

This can have serious consequences for public policy. Why is there still any kind of debate about the existence of anthropogenic climate change, long after the emergence of a consensus among scientists that it is indeed happening? One element of the explanation is that both sides spend a lot of time hanging out in online echo chambers. If you believe that man-made climate change is a scare story manufactured by the liberal elite, you will go to sites that confirm it is all a conspiracy. There is no way you'll end up on the Friends of the Earth website, for a different take. Similarly, if you're a paid-up member of Friends of the Earth, everything you read online won't necessarily make you believe that climate-change deniers are selfish rednecks who don't care a damn about the imminent apocalypse, but it certainly won't make you any more sympathetic towards their point of view.

When crucial issues get lost in the echo chamber, society suffers. One way to think of the bipartisan deadlock in United States politics is as a face-off between two clusters of people who've spent so long in the echo chamber that they are genuinely horrified that the other lot even exists. And in times of crisis, this polarisation is likely to get worse. Viewed from one angle, this looks dystopian. But of course, for us as individuals, life in our interconnected bubbles is simply more amusing and more convenient. A little bit more of what we like feels good to us.

That said, life isn't entirely libertarian in the heightened world of the echo chamber. Not when there are people willing to impose their own martial law on the wilder spaces of the Web.

37

HACKTIVISM

You think you already know what a computer hacker is: a vandal who breaks into websites to deface them or to steal data; or at the very least a perpetrator of mischief whose primary tool is his computer. That's been the popular understanding of the term since the early Eighties, but it's not the primary one. The term hacker has been around since the Sixties in techie circles and is used to describe someone who delights in pushing the capabilities of computer systems. In this sense, to hack something is to take it apart in order to be able to use it yourself in your own engineering. A hacker is one who can execute a particularly skilful piece of programming, someone who values creativity and collaborative effort in expanding the realms of the possible. To call someone a hacker is not to demonize them but to compliment them on both their skill and, here's the weird part, their integrity.

The general public may have images of pranksters in their mind when they think of hackers, but the truth is that those pranksters are rarely motivated principally by a desire to cause damage, or even the need to prove themselves. Their activities often embody what they think of as the ideals of the Internet. According to this point of view, companies like Microsoft and Apple, with their combination of closed data and litigious tendencies, represent not just a challenge, but an out-and-out insult.

The extraordinary sophistication of the IT world, and especially the Internet, would not have been possible without collaborative effort from a vast number of very talented people who put aside personal or commercial concerns and shared their ideas and expertise. That was true in the 1970s and remains true now. So corporations or government agencies that guard the secret of their engineering (not their content, mind you, but their engineering) are considered fair game by hackers.

Hackers share values and, increasingly, physical spaces. If you do a search for hack space in your local city, you will probably uncover a community of people who gather in a glorified shed to share machinery and expertise, to experiment and to teach each other how to be creative with networked technology. These are the people who are ensuring that the Internet continues to flourish in ways that do not simply benefit the big stakeholders. If we want there to be creative engineering outside of Silicon Valley, with aims other than to grow blue-chip businesses and enrich their shareholders, then we need hackers.

Hacktivism is the explicit pursuit of political activism via the prankster methods that characterise the tabloid press's typical hacker. There is undoubtedly crossover between hackers, whose motivation is primarily to expand the technological frontiers of the Internet, and hacktivists, whose motivations are many and various. For proof you need only look up Noisebridge, one of the most visible hackspaces. Their website announces that they are proud to provide IT services for the Occupy movement. (That's the thing about hackers: they like open data source, and sharing of information.)

Hacktivism is a huge and powerful weapon in the hands of numerous emerging pressure groups. 2011 was a year of unrest that was characterized in part by its techno savvy, and 2012 promises more of the same. But not all hacktivists are working to bring about the revolution. The website theyworkforyou.org was created by a group of people who considered it unacceptable that the record of the U.K. parliament, Hansard, was not available online in its entirety. When requests for full disclosure or access were rebuffed, the hacktivists "scraped" the official Hansard site for data and used it build their own resource for anyone who wants to know, for example, exactly what their MP said in a particular Commons debate. The site is now so widely valued that MPs themselves make use of it.

"Scraping" refers to the practice of mining a website for information that its administrators do not want to give you, and converting it into a readable, usable format for inclusion in your own work or someone else's. This is of course highly illegal, but, as with many other instances of the Internet's ethics coming into conflict with the letter of the law, the question is: should it be?

WikiLeaks is an extreme example of hacktivism, both in terms of the skill it took to pull off the various hacks, and the impact the site had on the off-line world. It pushed this debate beyond the realms of engineering geeks or the politically committed.

Ultimately, the fact that you *can* liberate information and share it with the world tends to produce that very result. It becomes not a possibility but a moral necessity. Hacktivists believe they have right on their side. And maybe they do.

38

HUMAN FLESH SEARCH ENGINES

The topic of this chapter certainly has some potentially alarming extrapolations, but the reason for its peculiar phrasing is that "human flesh search engines" is a literal translation from the Chinese. It was coined to describe what happens when an online community decides to take justice into its own hands.

In 2006 a video was posted to various Chinese file-sharing sites. It left a trail of disgusted and angry viewers in its wake. The video depicted a smartly dressed middle-aged woman stomping to death a small brown and white kitten with her silver stiletto heel. The video went viral and within hours there were calls for revenge circulating on online forums. The tone of the conversation quickly shifted to one of determined practicality. Users were urged to hack the video for clues as to its creator, and to pinpoint its location from myriad tiny identifiers. This communal detective work paid off (though it has to be said that the intervention of a popular newspaper accelerated the search) and the woman was named. Both she and the man who shot the film were sacked from their jobs and more or less run out of town. Online vigilantes secured their desired effect in the bricks-and-mortar world.

The human flesh search engine is particularly prevalent amongst Chinese Internet users, who have deployed it on targets from cheating husbands to corrupt local government officials. But this phenomenon is not confined to China—interestingly in the U.S., for example, the transgressions that have attracted the wrath of online communities have tended to offend against online etiquette rather than old moral codes. For example, there was the case of the fifteen-year-old blogger with terminal cancer who was revealed to be a perfectly healthy woman in her forties. That breach of the trust of the wider online community is different from the adulterous husband in China whose offence affected only his wife, friends and family.

As Internet users become more comfortable with changing their identity online, or forgoing it all together in the anonymous chat forums such as 4Chan, the potential both for fantasy and for bad behavior becomes ever greater. But the principles governing the retribution mechanism are the same whatever the offence: your misdemeanor is recorded and posted, so that it can be independently verified. When it is discovered, the online equivalent of a mob is assembled, often in less than twenty-four hours, and engages in a process that has become known as doxing.

Doxing is the matching of an individual's pseudonymous or anonymous online identity to their real-world one. If a dedicated group of twenty, or a hundred, or five thousand people is out to dox you, it won't take them more than a matter of hours, or minutes, to find your real name, address, employer's address and financial data.

It doesn't require a very vivid imagination to conjure the potentially catastrophic consequences of a vigilante mob turning up at the house of someone they have decided is in some way guilty. Part of the problem is that the human flesh search engine often swings into action not just against perpetrators of acts that mainstream society might agree were unsavory, but simply for the fun of it. You do not have to have done anything "wrong" in order to be doxed. The values of many online communities are mercurial, and essentially amoral. Kitten killers arouse their ire, but so do naïve teens recording painfully bad pop songs on their fifteenth birthday. The end result is a version of the same outrage or ridicule, and though the consequences vary in severity, they are similarly based in a desire to punish crimes against decency or taste.

An understanding of this capricious streak—and also, crucially, of the extent to which our lives are traceable online—is one of the issues that divides the Internet-aware from the unaware. If you are in the technically challenged camp, you probably find it frightening to be told that any halfway competent technology geek with a bee in his bonnet could find your home address in a matter of minutes and broadcast it to a baying online mob. If you are part of the technically savvy brigade, this news will come as no surprise whatsoever. In fact you've probably been using a fake address to register your domain names since 2003.

The problem is that online privacy for an individual is more or less impossible. Even if you decide to opt out of the game by, for example, not using social networks or shopping online, your data is still stored in numerous online databases, from Companies House to Tesco's. If someone with enough access wanted to, they could cross-reference all of these tiny snippets to assemble a very big picture indeed. And no matter how vigilant you are, you can be certain that your social circle and work colleagues are less so. You might never post your pictures on Flickr, but your friends probably have no such qualms. And if you're in those photos, you're potentially traceable by a human flesh search engine.

The flip side of the risk for the individual searchee is the safety in numbers afforded the searchers. For groups in societies that have very little access to any conventional form of social justice, the human flesh search engine is a tool for exacting it. Where corruption and disregard for the rule of law would expose a whistle-blowing individual to prosecution or worse, the human flesh search engine shields him. The confusing thing is that, as in so many of the phenomena we're looking at, the very same qualities that make this process laudable when the target is a corrupt official in a remote corner of China make it downright creepy when it's a teenager whose only crime is that she can't sing.

39

ANONYMOUS

If there's one group on the Internet that seems designed to make people feel old, it's Anonymous. Everything about Anonymous exemplifies the things that make the Internet different. They are a group with no geographical base but everywhere; no method for joining apart from to say you have; no central structure until they do, and then only for a few minutes; no aims, apart from when they're incredibly focussed; and no sense of seriousness, apart from when they're having fun. They are the Internet in its purest generation-gap role, and generally very misunderstood.

There is something very spooky but also very impressive about Anonymous. A group of anonymous individuals, they spontaneously organise to take collective action against a target, with results in both the virtual and the bricks-and-mortar world. Spooky, because they defy lots of our beliefs about how individual identity is constructed and the group structures required to get things done. Impressive, because of their self-confidence and status as an emerging powerhouse of rebellion.

Anonymous has developed into a movement whose action-movie-style mission statement—"We are Anonymous. We are legion. We do not forget, we do not forgive. Expect us."—is like a shot across the bows of the old order. And yet, it is not at heart a politically motivated movement (though its offshoots are heavily involved in protest) or a criminal one (at least not to any greater degree than hacktivism could be said to be criminal). But Anonymous will come looking for you if you offend against the Internet's code of ethics, and when it does, it is ruthless and extraordinarily powerful.

In the old days, if you wanted to organise an underground movement, by definition you were extremely concerned about the identity of the individuals you admitted to your confidence. Initiation tests, or at least background checks, would be required. You knew the names, or if not the *noms de guerre*, of your co-conspirators. You might be organized along collective or hierarchical lines, but there would be a clear power structure for decision-making and some kind of spokesman.

Anonymous is precisely the opposite. If you want to join, you just do. When you want to stop, you do that too. Nobody speaks on behalf of Anonymous, or everybody does.

It has its origins in 4Chan, an English-language version of Japanese message boards that facilitate a communal conversation about specific subjects. One particular message board on 4Chan—/b or "slash b"—has become infamous. There are no user accounts. You do not register your details in order to participate; you just post whatever you like, with no identifiers attached. The resulting slew of insults, schoolboy humour and soft porn has thrown up numerous classic Internet memes such as lolcats and Rickrolling.

Slash b is like the primordial soup of the Internet. It looks as if there are no rules, that the whole place has overdosed on the Online Disinhibition Effect, but actually it has developed not just its own in-jokes but also its own boundaries. It has, for example, been an enthusiastic adopter of the human flesh search engines that set out to punish cat abusers and other wrongdoers. (As an aside, there's something about cats on the Internet—cute pictures of them, videos of people abusing them. They're the gold currency of online chatrooms.)

Out of this environment there began to emerge a more self-conscious grouping with a sense of collective identity built around slogans such as "do it for the lulz" (fun) and "the Internet is serious business" (this being deeply sarcastic). With no signed-up members and no leaders, this was a group in the same way that a flock of birds is a group.

From time to time a chat stream on a message board would reach a critical pitch of resentment and a decision would emerge: to attack the offender. Unlike the objects of the human flesh search engines, these tended to be institutions, not anonymous individuals whose identities needed to be established. The first big Anonymous campaign was waged against the Church of Scientology. It was extremely effective, and it set the template for subsequent actions.

The Church of Scientology had made itself offensive to the incipient Anonymous group on 4Chan by demanding the removal of a video from YouTube. The video combined leaked footage of Tom Cruise speaking at an internal meeting and the *Mission Impossible* theme tune, and was declared to be in breach of the Church's copyright. It appeared at around the same time as Cruise made his notoriously odd appearance on *The Oprah Winfrey Show*. When the video remained live on YouTube, the Scientologists threatened to sue the platform.

This was not the first time that the Church had threatened action against its legion of online opponents. It's an organization that attracts a lot of accusations of human rights violations and cult-like behavior, many of which are expressed online. That video could have been made anywhere in the world, by someone who may or may not ever have heard of 4Chan. It didn't matter.

A furious conversation in the slash b forum solidified into a call to action. The group decided that the Scientologists needed to be taught that there would be redress for being so aggressively litigious against online entities. They launched Project Chanology. This combined hundreds of prank calls to the Church's headquarters, the sending of thousands of black faxes and a sustained denial of service attack.

A denial of service attack brings a website down by bombarding it with hundreds of requests per second to display a particular page. The site is overwhelmed and crashes. The programming skills required to write the code to attack the Church of Scientology's website were well within the reach of plenty of 4Chan's technology-literate users. The campaign succeeded in temporarily devastating the Church's ability to operate and established a precedent for using increasingly sophisticated software to hack other targets' websites. It also brought Anonymous to the attention of the world.

Since then the movement has grown online and off. Not long after Project Chanology, group members adopted the *V for Vendetta* Guy Fawkes mask as a visual identity and began gathering in the bricks-and-mortar world. (If you're ever on Tottenham Court Road in London on a Saturday morning, you'll see twenty to thirty members of Anonymous, all wearing masks, gathered on the other side of the road from the Church of Scientology in a silent protest.)

As the Scientologists discovered when they tried to fight back against Anonymous, it is extremely hard to counter attacks from a nebulous organization that doesn't have a leader you can appeal to or negotiate with, or even (at this point) its own website you can vandalise in return. An organization that understands the rules of the new online arena far better than you, has access to better programmers, greater numbers of people and that has no qualms about fighting by its own rules. An organization that has close to zero operating costs.

In the end, the Church of Scientology decided the best tactic was to ignore Anonymous's actions and wait for their attention to shift elsewhere. The Scientologists have grown markedly less litigious since they were targeted.

Meanwhile, Anonymous is increasingly blurring with a wider protest movement that in 2011 started to realize the extent of its power. Those iconic *V for Vendetta* masks were very visible in press coverage of the Occupy protests, for example. Mass popular movements, even where they are organized along far more traditional lines than Anonymous, can be incredibly powerful if they take advantage of the nimbleness of fluid networks rather than rigid hierarchies.

Francis Fukuyama's thesis about *The End of History and the Last Man* has been proved to be premature so many times since its publication in 1992 that it feels like a relic from a vanished world. The clash between the old hierarchical and the new networked generations is the latest manifestation of history's tendency to reassert itself. 2011 was the year that millions of people—not just the movers behind the Arab Spring but also the *indignados* in Spain and the Occupy movement—came to the conclusion that power in the hands of a few ruthless elites had got really out of hand and was screwing everything up, from their country's integrity to their employment prospects to the planet's ecology.

By contrast, for many of these (mostly) young people, all the good stuff was happening around them, in the networks of like-minded people they knew and respected. There is tremendous solidarity between these people. By way of example, Anonymous was crucial in launching denial of service attacks against the Mubarak regime's servers when the government was attempting to cut off the Egyptian people from contact with the world.

2012 will likely show the new protest groups' determination to embody a system of checks and balances to the old hierarchical power structures. One intriguing question is whether, as the economic and political situation gets worse, other more traditionally compliant groups will start to adopt the methods of the networked protesters. Just imagine what would happen if the members of the Countryside Alliance or Mumsnet took to staging the kind of nimble, focussed, savvy protests that the Anonymous movement have been perfecting. Hierarchies all over London must be dreading the day the baby boomers march on Whitehall in Guy Fawkes masks to demand the restoration of their pensions.

Anonymous are the Internet made flesh. Just as the Internet treats censorship as damage and simply works around it, so too do Anonymous. They are the physical manifestation of that social change, which has been brought about by the architecture of the miraculous World Wide Web. All of which makes them either utterly terrifying or completely thrilling, depending on your point of view. Spooky, certainly, but also very impressive.

40

IDENTITY PROVIDERS

It used to be the case that unless you were engaged in some exceptional activity such as writing subversive literature or spying for your government, you had just one identity, derived from your one name, given to you at birth by your parents and ratified by the State in the form of your birth certificate.

From this primary document of identification flow various others. Your passport is almost certainly the "strongest" piece of ID you possess. With a photo and your given name, as well as your date and place of birth, it's the trumps card of real-world ID. Driving licences, bank cards, your union or club membership card, all these various pieces of "lesser" ID are obtainable, so long as you have your passport, because the passport is the trusted ID, ratified by the ultimate identity provider. It follows that with different levels of identification you can do different things. It will take more and stronger forms of ID to secure a mortgage than it did to buy a martini back in your younger days.

We all instinctively understand that our names, and in fact our unique status as individuals, are solidified by the approval of the State. Our sense of self is of course a flexible thing, we may have gone by a nickname at school, have fluid social and business and sexual personae, but until very recently, none of those identities would be recognized by any entity beyond the specific group of associates with which we use it.

That all changed with the Internet. As I pointed out in the True Names chapter, you almost certainly have an email address, perhaps a Twitter name and a Facebook profile. You might have an ID you use to participate in online forums—several in fact. If you date online you have an ID, probably a pseudonym. If you play a multiplayer game such as World of Warcraft, you have an avatar. Even if your Internet use is restricted to reading newspapers behind a firewall and shopping on Amazon, you have identity tags for both those activities so that you can save your settings and keep records of your purchases. None of these online entities have asked to see your passport before you join up; they take your declaration of your name on trust.

And increasingly, services such as Facebook or Google+ are trusted by other "lesser" Internet entities, in the same way as the State is trusted by your bank. These mega sites now provide identities that are valid not just on their own portals but on numerous others. Facebook has become the online ID provider of last resort. Virtually everything you do online, apart from accessing your bank account, can be achieved without recourse to the name on your birth certificate. These online identities are just as functional, in many areas of life, as the names in our wallets.

There are plenty of reasons why we might want to avail ourselves of the freedom this functionality affords. They range from a desire to keep our private life separate from our work life, to a need to protect oneself from an abusive ex-partner, or to keep under the radar of a hostile political system. Pseudonyms are also helpful if we wish to dodge responsibility for something, but only temporarily—as we've seen, everything online is traceable in the end. More simply, multiple IDs for different areas of online life are now the norm for heavy users, simply because they have grown up on the platform and are accustomed to the flexibility and potential for reinvention it offers. If you are sixteen years old and going through a death-metal phase, it makes sense to you to tailor your online profiles accordingly. When you decide that Mongolian folk music is more your thing, you'll get a new name—simple as that.

But the use of pseudonyms online is not without controversy. If you check the terms and conditions of Facebook, you'll see that personal profiles should not be created for identities other than your given name. Technically, that means your full given name as it appears on your birth certificate, and nothing else. The reason for this policy is that Facebook makes its money by selling the data you generate when you use its site. In order for that business model to work to its maximum profitability, it needs "you" to be reliably you. For years Facebook has been turning a blind eye to the fact that, manifestly, there are millions of users who have multiple profiles with different IDs. But last year, they decided, presumably at the urging of their advertisers, to try to enforce their wallet-name-only policy. Except it's not so easy to do. Salman Rushdie was vocally disgruntled when Facebook altered his profile on the grounds that it did not meet these requirements. The author protested that no one had ever called him by his first name, Ahmed, even as a child, and that he was known not just to friends and family but to millions worldwide by his middle name, Salman. Facebook reinstated his profile.

Google got itself into a serious predicament when a row blew up over its decision to refuse Google+ accounts to anyone they suspected of using a pseudonym. One such person was an Australian artist, activist and digital technology specialist called Skud. The name on her birth certificate was Kirrily Robert but she had gone by the name Skud for years, using it in her working as well as her social life. She had even done some consulting work for Google and been paid through its payroll with the name Skud. Her argument was that Skud was her primary ID, given that she had been using it as such for years. Google+'s insistence on no pseudonyms was fundamentally backtracking on years of the natural and organic development in people's Internet use and now amounted to an attempt to strip users of their identity.

Google+ altered its policy. Its business model, like Facebook's, depends on selling access to information about its users, and it decided to chance its arm to see whether it could enforce the one-person one-ID policy that would make this business model even more solid. The fact that the U.S. Justice Department's battle against serious online fraud was having a knock-on effect on its attitude towards the use of pseudonyms probably also played a part in Google's initial policy. But in the end, the scale of people's attachment to their Internet-derived identities was too strong. Without the support of heavy users and early adopters, no mass-market online venture can hope to gain the necessary critical mass. Skud and others like her were too important for Google to alienate. Multiple IDs, pseudonyms and avatars are here to stay.

As always, society's ability to cope with the changes brought about by new technologies lags some way behind those technologies' development. There are pitfalls lurking as we try to figure out how to cope with, say, meeting someone at a social function whose blog we have read for years. We feel we know them, but in fact we know only that persona that blogs under the name TallulahTangoAddict. And if they know us at all, it's as the ID that leaves the occasional comment under the moniker BlueStilettos123. The question, "What should I call you here, in this setting, as opposed to in that other setting?" is one that more and more of us have to negotiate. Will multiple identities bleed together or should they remain distinct?

Although, it's not difficult to imagine that even if the Internet businesses settle upon a policy of tolerance, and we grow accustomed to multiple identities, there will still be instances in which their mishmash has to be disentangled and examined—if you are applying for a long-stay visa, for example, or a job. In such situations you have to provide a verifiable record of what you have done, and not done. How long before the authorities want to check not just your wallet name's criminal record, but also your online identities'? If I'm applying for a Green Card, it can't be long before BlueStilettos123 has just as much to prove as Ben Hammersley.

41

THE NICHE FACTOR

If you are thirty-something, you belong to the last generation of people who grew up with a dominant pop culture. It used to be true that one could identify the era in which a photo was taken from visual clues: the early Sixties looked appreciably different from the early Seventies. The folky hippy vibe of the early Seventies was wiped out by disco and punk. The early Eighties was all about New Romantics and then yuppies. By the time we got to the early Nineties it was minimalism and grunge. Obviously not everyone was a Mod, or a punk or a New Romantic, but the cultural mainstream was flavoured to a great extent by these dominant fashions. They were mass movements shaped by the forces of localism and the output of the entertainment industries.

If you went for a stroll on London's Oxford Street or New York's Fifth Avenue right now and took a selection of photos, I can guarantee you would be hard pressed to pinpoint a dominant aesthetic. This is not to claim that the Internet invented subcultures—people have always been drawn to the obscure, the cool and the marginal. And if you took your photos in, say, the streets of Harringay, home to London's Turkish and Kurdish population, or observed the Asian or Afro-Caribbean communities in Mile End or Southall or Brixton, you would of course get very different results.

Nonetheless, until recently there was a discernible mainstream culture with alternatives defined in relation to it. Now we are all cool-hunters. In the networked world we can find the other Peruvian nose-flute jazz enthusiasts in seconds. We'll probably discover that though we thought we were the only, lonely aficionado of a niche pursuit, there are in fact thousands of us, all over the world. So many that the niche needs its own niches. We can pal up with the other pre-1955 Peruvian nose-flute jazz diehards and ignore all the deluded fools who prefer post-1955 Peruvian nose-flute jazz. This represents either the fragmentation of the mainstream, or the mainstreaming of everything, or perhaps both.

The mainstream culture is not taking this assault on its hegemony lying down. Shows like *The X Factor*, *Strictly Come Dancing* and *Britain's Got Talent* are their last roll of the dice in the crap shoot with fragmentation. But even here, the influence of the new networked model is obvious. Take, for example, the fact that for the last three years a streetdance ensemble has finished in the top three on *Britain's Got Talent*. Streetdance is not a mainstream activity in Britain, not according to any conventional system of assessing things anyway. Streetdance is a niche activity, but it's a big niche. It's one of those pursuits that turn out to be infinitely more popular then you might imagine if you were simply taking the dominant culture's word for things. And when you grant a successful niche activity access to a truly mass-market platform such as prime-time television or the Internet, it assembles all its adherents and starts to punch at its true weight. Streetdance might not quite make a million-man niche, but knitting certainly does. So does rambling. And so does playing computer games.

Which brings us on to the other interesting thing. (Or scary thing, if your job is to prop up the dominant culture.) New forms of technology allow new forms of measurement of cultural status. Now that we buy online and even our high-street purchases generate accurate electronic sales figures, it is impossible to maintain that more people like something than is in fact the case.

The erosion of the authority of hierarchical systems of criticism has revealed different voices with different tastes. We no longer have to take Barry Norman's word for it when he tells us which is the best film of the week. The old custodians of mainstream culture, whether highbrow or defiantly popularist, have tended to be members of the cultural elite. The commissioning editors at broadcasters and publishing houses and advertising agencies are almost uniformly white, middle class, university educated etc., and most of them are instinctively wedded to the hierarchical way of doing things. With relatively few exceptions, they tend to react with panic or a sort of naïve enthusiasm to the revelation of previously undreamt of niche markets. It's not their fault—they're not the right people for the job of delivering culture to those niches. The fact is, now every seventy-five-year-old streetdancer from Blackpool or nineteen-year-old knitter from Tottenham can access and create their own niche culture.

Our cultural industries remain in thrall to patterns of consumption that are hopelessly out of date. That is proved by the single fact that there is virtually no discussion of computer games as cultural products anywhere in the mainstream press, despite the fact that they are now as sophisticated as Hollywood films and far more people play computer games than play football.

Football should be a victim of the new accounting. It has a disproportionate amount of attention paid to it, considering the numbers of people who actually watch live games, or even *Match of the Day*. In fact, we now persistently miscalibrate the numbers that justify taking notice of something, both by overestimating and underestimating. In the old days, the BBC considered 5,000 letters from readers to be evidence of a very strong level of approval (or disapproval). These days the Marillion (Eighties prog rock, for those who weren't there) fan site can generate an email with 20,000 signatures within twenty-four hours if it feels the need. We must develop a new scale for measuring people's levels of interest. And we need to accept that it might not tally with what our cultural instincts tell us. This could be very exciting: it might lead to the discovery that there is a superstar Turkish musician who pulls in crowds of 10,000 night after night at venues all across North London, and that we love what he does. Or it might prove to us once and for all that the world is a baffling place full of people who do not think as we do. Whatever the case, one thing's certain: we're in the midst of an epic cultural revolution, and the mainstream media can't afford to ignore it.

42

DIGITAL RIGHTS AND WRONGS

The question of protecting content against piracy causes a huge amount of misunderstanding and bad feeling on both sides of the debate. Every two or three years, old-school content providers such as newspapers, book publishers and record companies unveil a new digital rights mechanism that will control access to their products, and thereby hopefully generate more income from them. This is of course perfectly understandable—their business model demands that they should. And again, no one (least of all me!) would argue that writers, editors, musicians and all the rest do not deserve remuneration for their efforts. However, relying on technological means to lock content to one single purchaser, rendering it incapable of being pirated, is utterly pointless. Let's examine why.

The key issue for content producers is that digital files can be copied perfectly and without cost. If you have an unlocked version of this book as a PDF file, you can copy it an infinite amount of times, and every copy with be identical to the first. In technical terms this is not the same as copying a cassette tape, or recording something off the radio. Furthermore, if you were to put a copy of this book onto a website, you could make it available for free to millions of people for negligible cost. My publisher would be upset with this, as are their brethren in other content publishing industries. The fashion, then, is to treat the file with any number of Digital Rights Management tricks, to tie the copy of the book you have bought (thank you, by the way) to you the purchaser, rendering it unreadable by anyone else. You can copy the file as much as you like, but only the software it was specifically bought for can read it. Online music stores, streaming video, and other such content use similar methods.

These Digital Rights Management techniques are a form of encryption, where only you the purchaser have the key. The math behind the DRM techniques are irrelevant. It doesn't actually matter just how secure the file actually is, because at some point, the computer has to decrypt it for you to be able to enjoy the content. Once it is decrypted, and being played out of some speakers, displayed on a screen, or whatever, it is no longer under any form of protection. Even the most sophisticated piece of music DRM is defeated by a 50p piece of cable linking the earphone socket on one machine to the microphone socket on another. And once a single copy of the content is available online without any DRM, the whole DRM regime is broken.

This is an important point. There are many ways for the piracy marketplace to gain a copy of something without DRM—it might be copied as above, it might be leaked from the studio, it might be recorded by someone with a video camera sat in a cinema—but whichever way it appears online without any restrictions, once it is there, any attempt to enforce DRM is already doomed. Publishers need to find new ways of operating in an environment where it is certain (or perhaps, hopeful), that people will be making their content available for free to those who know where to look.

Another restriction technique, though not specifically DRM based, is one of geo-blocking the content. This is where a publisher makes something freely available only to users accessing the Internet in specific countries. BBC iPlayer is restricted to the U.K., for example, and Comedy Central is geo-blocked outside of the U.S. But with a Virtual Private Network, an encrypted connection that hides your actual location (available at a fee of just a few pounds a month to anyone anywhere in the world), you can trick any service into thinking you are in the country of your choice. I live in the U.K., and pay my BBC licence fee very happily, but when I'm in the U.S. I can't watch shows on iPlayer. Until, that is, I connect to the Internet via a VPN whose servers are in the U.K. To the rest of the Internet thereafter it appears I'm in Guildford, and not San Francisco, and iPlayer works as at home. Services technically restricted to Americans are easily accessible in the U.K. by using the same method in reverse. Geo-blocking is trivially easy to circumvent.

This is not just for online content. Take the case of fan translations. Because book publishers sell rights to different language versions, and because, say, the German edition of a Harry Potter title might lag six or more months behind the English one, German Harry Potter fans had to wait to get their hands on a copy. Or rather, they officially had to wait. Internet savvy fans invented a way to get round this problem, and it beautifully replicates the basic mechanism of the Internet's own information delivery system. Someone bought a copy of the English editions on their day of publication, sliced the covers off, scanned every single page and posted it to a fan site. Then a crack team of perhaps 400 English-speaking German fans got to work translating a couple of pages each. When they were done they posted their translations and hey presto, perhaps twenty-four hours later there was a workable German translation available for every single fan. For free. Now it's true that the Harry Auf Deutsch project was shut down by the German publisher threatening legal action, that they stopped is perhaps more to the credit of the law-abiding fantasy fiction fan than anything else. These are people who would probably have been happy to pay whatever the asking price for a digital version of the book, had it been available. That is wasn't could be considered a market failure. And they do not consider it in any way wrong to have retaliated against the unreasonable restrictions placed on their desire to buy, read and join the global conversation about the book at the time it was released in English.

All of which illustrates the point that restricting access to digital content doesn't mean that people won't be able to get hold of it; it simply means that they won't have to pay you for it when they do. (This might be another of those counter-intuitive moments that presents problems for people who are not familiar with the basics of why the architecture of the Internet makes this a simple matter of truth, not a matter for moral debate.)

What this means is that it is increasingly difficult to protect national or commercial interests online unless you provide a compelling positive reason for people to sign up or pay up. In point of fact, people will pay for the content they access if they value it enough. But you, the provider, have to make them value it sufficiently; if you tell them they can't have it, they will simply go elsewhere, and you will lose the revenue. This is borne out by the success of the Apple store. The vast majority of the music available there for purchase is free elsewhere, but consumers value the ease of access, additional information and customer service. They also like being able to pay the producers of the music—something that handwringers and naysayers all too often forget in their haste to defend digital rights against the scourge of the Internet.

43

THE FUTURE OF MEDIA

There has been more angst about the slow death of the old media than about any other sector that has felt the creative, destructive power of the Internet. That's not surprising: it is the job of journalists to make a lot of noise, and when the subject is their own livelihoods they're especially vocal. I'm a journalist myself, so I have both affection and respect for the business. I believe it's important that we have good quality journalism, and that writers' books are edited, and that television covers current affairs. I also believe that many outlets of the professionally produced media will flourish, even as the entire industry shifts online. That might sound unlikely, given what we're constantly told about the public's refusal to pay for any digital content and the assertion that the Kindle will destroy books, that there will be nothing but unformed opinion in the place of newspapers etc. etc., but I would argue that the benefits of the shift to digital are far greater than its drawbacks for almost everyone, consumers and producers alike.

Firstly, it would be facile to deny that digital's arrival has caused a great deal of turmoil for newspapers, magazines, television companies and book publishers, turmoil that they have succeeded in relaying to their customers, many of whom are also concerned. Unless you are actively involved in digital media, it can be easier to feel the destructive force of the Internet's cyclone effect than the exhilaration of its creative impact. When local newspapers are all but dead, national newspapers are under threat and commercial television is nothing but talent shows, it's easy to worry that professional media made by skilled and experienced people is finished, and we will be left with nothing but a morass of nonsense on the Internet.

In order to understand why we should not despair, it's helpful to understand exactly what's at risk, and what isn't. Professionally produced quality media is not going to disappear, but the one-size-fits-all business model that has served since the beginnings of mass-media consumption will. In fact, it has already died.

The fundamental purpose of all commercial media, whether it's ITV or the *Daily Telegraph*, is to make content that attracts an audience. That audience is then sold on to advertisers. If you pay for the media you consume, as you do when you buy a printed copy of the *Daily Telegraph*, you are providing another revenue stream for its producer. Your contribution is nowhere near as important as the advertising revenue mind, but it's good to have. If you have not paid, as is the case when you watch ITV, then you are the only product. In either case, what the advertiser is paying for is your attention.

This all worked fine until the audience for newspapers and television programmes started to fragment. It wasn't that people no longer wanted to be informed about the world, or to be entertained, it was simply that they had so many more options. More TV channels; more books being published; new media such as computer games; and of course the prize villain: the Internet. The Internet provided an embarrassment of options for consumers, but crucially it provided most of them for free.

Local newspapers were the first to sink under the strain. Individual components of their revenue stream migrated to the Internet in quick succession, starting with job advertising, via holiday adverts and then the classifieds. As we saw in the chapter on asymmetry, it was Craigslist, the free online local classified service, that really sounded the papers' death knell back in the 1990s. Why would anyone bother to pay for an advert in a single newspaper when they could post one online for free that gave infinitely wider reach? The answer was that they wouldn't. Craigslist had broken the economics of regional newspapers, and the implications were ruinous for every other mass-media channel.

Since then, the "Craigslist Effect" has meant a dramatic escalation of the forces for change. Advertisers increasingly prefer online to print or TV advertising since it allows them to target their audience and to collect accurate data about them that they can use to improve their strategies. Even if the media companies can shift their business online and take their readers with them, there are no easy answers when they got there. Conventional wisdom says that all content had to be free on the Internet, so there's no secondary revenue from the cover price any more. And the advertising revenue has shrunk considerably thanks to the fragmentation of the audience. In the face of cuts to editorial budgets and job losses, it's easy for media workers to believe the end is nigh and to tell that story loudly to their audience.

But here come the reasons not to despair. For a start, if we are honest, we can probably admit that much of the old media was churning out pretty mediocre product. Now that the economics of online advertising require a shift either upscale to a quality niche market or down to the still very substantial mass market, a great deal of middle ground will indeed disappear. But even so, the picture is not all bleak. There is plenty of thriving professional media online that have retained elements of the old way of doing things and capitalized on the creative powers of the new platform.

If your concern is that quality media will not survive, you have scant reason to worry. The *London Review of Books* and the *Financial Times* are examples of successful publications that prioritize the quality of their coverage and writing. They have a valued product to sell and have adjusted well to the ongoing transition to digital. Both give away a small amount of content for free on their websites and charge for access to the rest. Both employ a model of membership to a network of like-minded people, rather than old-fashioned subscription, and both have realized that the Internet allows them to sell additional services. When you pay for content via either of their sites you are also buying access to events and conferences—to an association, essentially. The value of the product is clear, subscriptions hold steady or increase and advertisers continue to pay for that particular consumer's attention.

Similarly, some mass-market products are holding up very well: the *Daily Mail*'s website for example, is the most visited English language newspaper site in the world. Here, again, the readership and advertisers are still connected to one another in the same way as they always have been, via the content. The *Daily Mail* does not charge, but it doesn't need to, thanks to the economies of scale.

It might be more accurate to say that the old business model, rather than being broken, has grown better at delivering what people actually want, whether that's extended essays on the legacy of fascism in post-war Italy or celebrity fashion and gossip. And the panicky truism that no one will pay for online media turns out not to be true at all. When people value content they will pay for it, especially if its producers use the group-forming and networking powers of the Internet imaginatively to offer extra services.

This is the case for music and books as well as periodicals. Yes, anything in a digital format can and will be downloaded for free, but if you give people a reason and the opportunity to buy your product, many of them will. You might need to find additional ways to make money to compensate for lost sales (live performance, say—and the growing number of literary festivals gives the lie to the idea that people won't pay to see writers perform) or you might find that actually, a digital platform allows you to reach new readers and grow your market. This is the reason that Kindle will not in fact destroy book publishing. The number of physical books sold may fall, (though there is no reason to suppose that they will no longer be produced) but the number of unit sales will in fact increase if existing trends continue.

The flip side to the hollowing out of the old professional media is the explosion of increasingly confident and assured amateur journalism and creative work. The fact that anyone with a smartphone can make videos enables revolutionary protest and citizen journalism alike. Access to media technology and free distribution have democratized the production and consumption of information of all kinds. Freedom of the press now extends far further than ever in history, way beyond those who own the presses. There are now literary or food bloggers, say, who are courted by the publishers and restaurant PRs who used only to talk to a handful of columnists on the national press.

Not all of these amateur writers, film-makers and critics are producing "good" work, but some of them are. After all, you might know a great deal about cooking and be able to write a fine piece about a restaurant without having undertaken a journalism course or been paid for your reviews. The new breed of media personnel are learning their crafts through the mass experiment that is the Internet. Some of them are transforming themselves into professionals by sheer dint of actually selling their work themselves, without the permission or approval of the old media authorities.

That presents an uncomfortable truth for the former gatekeepers and taste-makers. They are now in competition for the public's attention and money with everything on the Internet, which increasingly means everything that has ever been produced: videos of kittens; the entire works of Anthony Trollope. The rules are the same for all of us now: if you're making work that people really want, high- or lowbrow, money can be made. If not . . .

That's terrifying if your job depends on a media corporation that hasn't adjusted to the new reality but is extremely exciting if you're an aspiring restaurant critic or you've written an alternative guidebook to the tango subculture of Buenos Aires or you're drumming up an audience for your first documentary film. And even if you're a professional at that struggling corporation, you will almost certainly find that your skills will be in demand elsewhere, so long as you can appreciate the creative rather than the destructive force of the Internet and learn its new ways.

The professional media is still with us, and now so is an army of passionate amateurs who are making and selling their own work. The lines between the two, and between consumers and producers, are blurring as the Internet remakes an industry we all have a stake in.

44

REMIX CULTURE VS. COPYRIGHT LAW

When culture is online, at the most basic level it's not the last installment of Harry Potter as an ebook, or a Tchaikovsky violin concerto in MP3, it's just data, which means that it can be manipulated. With a personal computer or a smartphone, we are all capable of writing our own alternative version of Harry's showdown with Voldemort or splicing Tchaikovsky with the Stone Roses, and then releasing the new version to the public. The results might well be banal, or unlistenable, but the technology means it isn't difficult to do.

Even if our participation in remix culture extends no further than touching up our summer holiday photos, we are all increasingly familiar with the idea of being able to manipulate what we see or hear on-screen. People who already live their lives online absolutely demand it. As we've seen in the chapters on hacktivism and Anonymous, creating open access in order to make creative use of online content is the cornerstone of the Internet's emerging moral code.

Cultural artefacts have had a contradictory status for hundreds of years. Stories, tunes and characters are both the singular creations of their authors but also tropes in our common culture. The balance of ownership has been shifting in the individual creator's favour virtually since the world's first copyright statute was passed by the new U.K. parliament in 1709. A twenty-first-century definition is very clear: Harry Potter is first and foremost owned by J.K. Rowling, her publisher and the production company that makes the films. It is true that Harry is also a meme in the world, with an existence outside of the content business, happily living and evolving in the heads of his fans, but while this is acknowledged, it is a truth that, for Harry's owners, requires managing.

Actually, Harry Potter is also really just data. Which means that many people (millions of them, since the core Harry Potter audience is precisely the first generation that has grown up online) have an instinctive belief that they are, for example, entitled to write their own versions of the stories and share them.

This is not because Harry Potter fans wish to trample all over the concept of copyright for their own amusement, or to compete with the mainstream content industry. It is simply that the human instinct to make more art out of whatever is lying around in the cultural common realm has coincided with a revolutionary tool for creativity, and crucially, publication.

Naturally, many copyright holders are of the opinion that where remix culture is in conflict with copyright law, the law should be upheld. But while this may appear self-evident to them, to many in the networked generations the opposite point of view also feels like a self-evident truth. This amounts to yet another gulf between the technologically illiterate hierarchical generations and the technologically savvy networked ones. After all, what is the worth of a creation if no one pays it any attention? But does paying attention to a creative work inevitably lead to that creative work living a life of its own?

The movement for copyright reform is growing. That movement has tended to be characterized by the mainstream media, which obviously has a vested interest, as wantonly indifferent to the law and to the right of creative people to be paid for their work. It's fair to say that neither of these things is true.

Firstly, it's worth remembering that laws are, like everything dreamt up by human beings, products of their own time. Copyright law didn't exist until societies in the process of shifting towards the recognizably modern world needed them, to encourage innovation and foster creativity. The Enlightenment ideals of the individual's worth were then refined by Romantic philosophies and resulted in an emphasis on the exceptional, the lone genius producing his masterpiece, whether in engineering or literature. But the original copyright legislation acknowledged that in reality there is no such thing as the genius working in isolation. In fact, copyright represents an agreement that, since society wants new and exciting work to happen, it will grant the authors of that work a temporary monopoly on its exploitation. Far from enshrining a concept of ownership that derives from originality, copyright says that society has a stake in your output, because your capacity to create it is dependent on input from the wider culture. The quid pro quo is the eventual return of ownership to the public domain, so that your idea becomes part of the common realm for other people to use in turn.

The length of the term of copyright in the U.K. was originally fourteen years from the moment of the work's creation. Today, for print works at least, it is seventy years from the death of the author. The first term allows an individual ample time to make some money out of his creation. The current term enshrines the right of perhaps three generations of inheritors, people who had absolutely no input whatsoever to the original work, to exploit its commercial value. Or to do nothing at all with that work except insist that no one else can use it.

It's clear that the original social contract represented by the idea of copyright has been completely broken. Or, rather, that it has been modified repeatedly over the years in the interests not just of individual artists or engineers and their dependents, but whole supporting industries of people. Copyright exists now primarily to protect the mainstream content-producing corporations.

Disney, for example, lobbies constantly to extend the copyright terms in order to prevent characters such as Mickey Mouse from falling out of copyright. Its efforts are consistently successful.

This ferocious emphasis on ownership by individuals and corporations has always been in tension with the instinct to create freely, but as we've seen already, when power lies in the hands of a few broadcasters or publishers there is very little check on their determination to protect their interests. That all changed with the shift in power brought in by the Internet. Now the pressure on copyright law is at crisis point. When new technologies (or new beliefs, or new whatever it is) arrive to alter the texture of the world, laws need to be flexible enough to change. If not, when they slip too far from a social contract that respects differing rights, inevitably there is conflict.

The suggestion that copyright reformers don't care about creative people being rewarded is unfair. Many of the people campaigning for more flexible copyright are themselves creatives, but from the "open source" school of thought. They maintain that it is perfectly possible for writers and artists and engineers and craftsmen to make a living from their art. And in fact, so can many of the skilled middlemen or suppliers of the creative industries: the editors and sound engineers and producers. But the hierarchical structures that ring-fence those people are a form of protectionism that squashes creativity and stifles innovation in favour of protecting corporate interests.

Part of the problem is that mainstream society has grown so used to a baby-boomer generation's definition of originality. The same people who sneer at remix culture for being unskilled splicings or shameless stealing are at pains to insist their own work is pure genius, created in whole cloth, even if it is demonstrably, say, a fusion of Delta Blues and early rock and roll. In this reading, originality was invented in 1963 (or at least between the end of the Chatterley ban, and the Beatles' first LP), and must be protected for evermore, or at least for long enough to give three generations a comfortable ride.

This is not the only truth about originality or creativity. The musical innovators who in the past six or seven years have created mash-ups (joyous, startling blends of unlikely but harmonious music) are making creative fusions every bit as original as the work of the pioneers of guitar-based rock. And actually, while we've known since the punk movement that anyone can play guitar, the digital instruments of choice for this new generation of musicians demands a level of skill that require far more years of practice.

While it seems that 95% of the content of YouTube falls into the category of "*Match of the Day* clip, set to the Beastie Boys," that means that the other 5% is good work. These proportions are much the same in every field of endeavour. We may not think much of the Harry Potter erotic fan fiction in literary terms, but that isn't really the most interesting thing about it. The fact that it is written, shared and enjoyed by hundreds of thousands of people demystifies the notion of the lone genius. Most creative work is derivative, of necessity. The artist works not in isolation but in networks with other people, appropriating bits and pieces and making something new. Sometimes it's good, often it's not, but the process itself can be fun and rewarding for its own sake, and not merely because its product has commercial value to the content industry.

Are we comfortable with the idea that an author of some of, for example, the most artistically brilliant mash-ups online, could be prosecuted for making her art? Do we believe that the owners of her source materials have rights that society must defend, even to the point of custodial sentences for a new generation of artists? At the time of writing, the attempt to pass draconian legislation in the States that would jail such people has just been blocked, and the bill postponed. SOPA (the Stop Online Piracy Act) would have allowed five-year sentences to be handed out for individuals breaking copyright law. We need a much better informed debate about which of two contradictory ways of thinking about creativity is more important to us now. It's time to follow the arguments to their logical conclusions, because in reality the change has already come.

45

OPEN COURSEWARE AND COLLABORATIVE LEARNING

Aside from all its other functions (as shop front, meeting place, crucible of creativity) the Internet is one vast educational tool. It's ideal for autodidacts, and lends itself perfectly to interactive learning via communities of interest. It provides exceptionally rich teaching materials and simple, cheap and effective ways of linking students to one another and to teachers, free of the time and location constraints imposed by attendance at a physical college. The Internet is a dream for anyone committed to securing or providing an education.

It also places considerable strain on the viability of the traditional hierarchical structures of the education system, particularly higher education. At the time of writing, students in much of the developed world are questioning whether leaving home to undertake the classic three-year degree is economically realistic or useful. Politicians and education providers are up against the limits of viable expansion of the university sector. Everywhere people are asking themselves, "What is education for?" Open courseware and collaborative learning online are suggesting new approaches to answering that question.

There are practically limitless educational resources online, and only some of them are provided by traditional institutions like universities. The Khan Academy, for example, a private initiative, has a series of over 3,000 video tutorials on subjects ranging from the history of western art to calculus, all of them uploaded by volunteers and free to access. The TED Talks project provides more than 900 video lectures from experts in their fields, on subjects ranging from cooking as alchemy to whether astronomers can help doctors improve their practice.

The driving passion behind these ventures is a classic liberal belief in the power of a good idea to change the world, coupled to a very twenty-first-century commitment to open access and huge excitement about the reach offered by the Internet. Even the ultra-prestigious bastions of hierarchical learning such as Harvard and MIT have made whole courses available online through open courseware. You no longer need to gain entrance to MIT to follow its Introduction to Psychology or Aerodynamics of Viscous Fluids courses. With lecture notes, reading lists and worked assignments, you can study from the comfort of your own home, to complement your accredited education, or simply for interest's sake. Other institutions across the world have used the MIT open courseware to redesign their own courses to make them more student-focussed or up to date. Access to such a resource seems faintly extraordinary to anyone whose experience of education is based on the fundamental principle of excluding more and more people at every consecutive stage. Education is a pyramid, with Harvard PhDs at the top. Or at least, it used to be.

The rejection of the old way of doing things, embraced by some of the providers and many many learners all over the world, raises all sorts of questions. If you can learn about viscous fluids with some of the best teachers in the world, for free, why would you pay up to £10,000 a year to go anywhere else?

Some answers are immediately obvious: for instance, to get credits for having done the course. MIT is not going to credit you for your online learning. You cannot claim to have attended MIT simply because you followed various of its online courses. Another factor, particularly in those countries where traditionally higher education involves leaving home for the first time, is the social and life experience. And there are many other reasons to believe that education in the flesh-and-blood world should remain the primary goal of both providers and students. Maintaining a flourishing ecosystem of institutions, for one. Not every university needs or wants to be Harvard.

Having said that, change is inevitable. Both the accreditation function of a university and its status as a social hub could easily be taken on by other entities. It's possible to imagine a central body running examinations for students who had followed a selection of different courses from different institutions. If that sounds like chaos to you, be assured that it is ever more likely. And as financial pressures make leaving home to attend university less and less viable, and social life shifts online, the student bar will diminish in importance.

In some subject disciplines and economic areas the landscape has already changed completely. A degree has traditionally been your pass to the first rung of a hierarchical career structure, but its value is increasingly questioned, especially by those who are entering the many digital industries that didn't exist even five years ago. A degree seems a convoluted way to prove you have the skills to do these jobs. Online skill-based accreditation systems are now emerging which are more transparent and fit for purpose than the standard undergraduate degree. Especially in an increasingly globalized world where beyond the big education brands, an A-level or a BSc might mean very little to a potential employer in another country.

At their most radical, these credits derive from peer-to-peer assessment on collaborative learning sites. Stack Overflow is a collaboratively edited question and answer site for computer programmers. When you participate in constructive ways, you gain badges and increase your reputation. There are badges for asking notable questions, for correcting your own work in response to others' input, for resolving others' problems. The site has a careers section that connects jobseekers and employers. If you wanted to recruit a programmer, why would you ask them whether they had a degree when you could simply look up their profile on Stack Overflow? The Khan Academy uses similar processes of gamification, Quantified Self and peer-to-peer review to award credits to its learners across all disciplines.

For now there is still a great deal of prestige attached to some institutions and some courses that only run in the off-line world, and of course the fact that MIT's open courseware is so well received and well regarded derives from its real world reputation. But the fact is, reputation for excellence is the key quality in MIT's status, and such markers of excellence are increasingly derived from a whole range of sources.

The question "What is education for?" is a profound one. But some of the easier answers—to get you a piece of paper that leads to a job—are under more and more pressure from social, economic and, increasingly, digital factors. In the future there will be many more sorts of learning. And meanwhile, the Internet is already a treasure-trove of all human knowledge, just waiting to be explored.

46

GAMIFICATION

You may never have played a computer game in your life, but if you use the Internet at all you are in contact with game design every day. The gamification of our world is spreading beyond cyberspace and turning virtually every area of human activity into a game. We're all players now.

The reason this is so, is simple. People, lots and lots of them, like playing games, the more complex, interactive and stylish the better. And the design of these games is now a creative industry to rival Hollywood in its heyday for technical wizardry, magic spinning, revenue generation and serious cultural reach. They are a major cultural and economic force, and if you're one of the people not aware of this, it can be deeply surprising to discover the sophistication of the art. The purpose of the latest generation of computer games is obviously to entertain its public, yes, but the way it achieves that is by influencing them to play by the rules and behave as the designer needs them to for the game to work and to make it entertaining. In other words, games make it fun to learn to do hard things.

Game designers have invented endless ways to make their products compelling and to coax their players in particular directions, drawing on psychological research as well as artistic vision. Game players like to be rewarded for achievement. They like to have a personal score they can set out to beat, as well as participating in challenges against another person. They love to accumulate points that can be redeemed for rewards. Game players like a challenge that is just difficult enough, but not impossible. They enjoy learning new skills and having those skills recognized.

This is called gamification, and its influence over behavior is extremely interesting, obviously, to corporations' marketing departments, but also to politicians and social scientists. All of these techniques for getting someone to perform certain actions are now used in the design of other apps and websites and are being pressed into service to encourage people to lose weight, share their expertise or buy a particular brand of chocolate.

Facebook's Friends count and Twitter's prominently displayed number of followers set an implicit challenge to compete with other users. Many people respond to this by increasing their activity levels and the upward spiral of use drives the entire machine of social networking. Foursquare—slogan: "Make the real world easier to use"—rewards users for logging their presence in various locations with a series of themed badges. There's a Swarm badge for checking in to a place where a hundred other Foursquare users are already logged. Or a Drinking on a Schoolnight badge for checking in to a bar on a weekday. Getting a badge is a sort of low-grade thrill that keeps you coming back for more.

These are fun devices that boost commercial products, but increasingly it's becoming possible to imagine more far-reaching effects being brought about through the application of game psychology. WeightWatchers, for example, has shifted much of its activity online. No longer a diet as such, their new strap-line is "The Game You Play To Lose Weight."

There's no doubt that certain behaviors can be engineered in the real world as in a computer game, but the design of game elements often implies that a person's motivation for behaving in a particular way is simple. That sometimes turns out not to be the case, particularly in more nuanced and emotive situations.

Research has suggested that applying gamification techniques to encourage people to volunteer more, for example, may actually be counter-productive. One can imagine a system in which voluntary activity in the community was rewarded with public recognition, with a sliding scale of "points" being earned, depending on the degree of work undertaken. But actually people's motivation when they undertake voluntary work is too subtle to make this a reliable way to influence their behavior. Some people might commit to volunteering if they thought their local council would publish their photo in the local newspaper, but at least as many people would refuse to participate in a scheme that rewarded their instinct for public recognition rather than their altruism. Perhaps not all areas of life are susceptible to game psychology after all.

47

DIGITAL AND ALTERNATIVE CURRENCIES

In the online universe, everything we encounter is a coded representation of itself. This has had the effect of making the tokens of exchange that surround us in the flesh-and-blood-world even more abstract. Ticketing, for example, is no longer reliant on a printed representation of the exchange of cash for access to, say, an aeroplane. Nobody has needed a physical ticket to board a flight for years, simply a booking reference and your boarding pass might be an image on your iPhone. The world has of course been accustomed to dealing in tokens to facilitate commercial exchanges for thousands of years, but these days, as with our plane tickets, there are abstractions piled on abstractions.

Money has been growing increasingly abstract ever since gold gave way to promissory notes and thence to cheques and credit cards. In the latter years of the twentieth century all sorts of pseudo-currencies appeared, from bonus points earned on your credit card to air miles. Now, with e-commerce, you can spend your money without it having to transform itself even temporarily into the physical tokens you can hand over to a shopkeeper. Money (like everything else on the Internet) is merely a string of code, zapped from bank account to bank account.

The ongoing global economic meltdown has exposed the modern financial system as an abstraction so abstract that it needs artificial intelligence to control its machinations. There is a growing longing to return to a simpler, more reliable sense of exchange. Something along the lines of, "I give you these coins in return for those potatoes." Or these online tokens in return for that downloadable music file. You see it in the multi-user online games, such as World of Warcraft, where you can trade or win gold, but speculating with it will get you nowhere. (Though incidentally, there are now online bureau de change just for game currencies, so that if you decide to switch from playing World of Warcraft to Final Fantasy, you can change your WoW Gold into FF Gil.)

One of the most concrete tools for rethinking our increasingly discredited mode of capitalism is an alternative currency, to be used either in a specific, local area of the off-line world, or solely online. This is not a new idea: community currencies have been in use in the States and Canada since the early 1990s, and at the turn of the twentieth century some companies paid their workers in tokens that could only be spent at the company store. But the appeal of these alternative currencies has never been so widespread as it is at the time of writing.

In the U.K., local currencies have been adopted in Totnes, Stroud, Lewes and Brixton. Bristol's scheme is launching in 2012. They have typically been promoted by groups with a broad anti-corporate or anti-consumerist ethos. The idea is simple: a currency that can only be used in a restricted geographical area, backed by deposits in sterling and distributed and regulated by a not-for-profit organization with the aim of supporting local businesses and the local community. Their driving principle is that when transactions are carried out in Brixton pounds, say, with an independent trader, the value of the money continues to circulate in the community, unlike pounds sterling, where on average only 10–15% of the value of any transaction stays in the area. The rest flows from Tesco's to its shareholders, on to the global financial market and ultimately into the pockets of a tiny minority. The schemes represent a return to an ancient notion of money, pre-credit default swap mechanisms, pre-futures derivatives, pre-bank loans, even, since you cannot bank or loan this currency, only trade face to face with it.

But if this is all sounding pretty Luddite for a book about future technologies, the really clever innovation of the new wave of local currencies is the ability to pay via text message using e-currency. Since digital retail is increasingly the dominant platform for shopping, the local currencies would be massively disadvantaged if they did not allow digital transactions. Thus their backers have developed software that allows you to send a message to, say, BrixtonPound.org, authorising a payment of B£6.53 from your online BrixtonPound account to Jane's Fruit and Veg Stall. BrixtonPound then texts both you and Jane to confirm your payment, in as little as twenty seconds.

These local currencies are a complement to pounds sterling, not a replacement, but they have proved successful and popular. Their advocates point out that these days it makes more sense to trust your money to a local credit union formed by known individuals, than to trust the algorithm machine that "runs" the global economy.

The BrixtonPound successfully combines physical and e-money, but the race to invent a purely digital currency with no backing from a state-issued currency has been a long-cherished project of digital enthusiasts. Motivated partly by a typically geeky desire to push technical boundaries and partly by an also typical preference for networked authority, the projects had always foundered on the same logistical problems. It was impossible to devise a system that was trustworthy and functional, that had some collective authority to prevent people from spending the same "bitdollar" over and over again. The various digital currencies never really amounted to anything more than counterfeit money for a handful of enthusiasts, until January 2009, when out of nowhere, it looked as if finally someone had succeeded in inventing a genuinely new form of money.

Bitcoin was, briefly, a fundamentally alternative exchange system, independent of any state-sponsored money market. You might choose Bitcoins because you were engaged in clandestine activity such as selling prescription drugs online, or, as with the founders of the BrixtonPound, simply because your personal ethics were offended by modern money. Either way, the subversive power of being able to trade outside conventional systems was hugely appealing. It was also too good to be true. Within a year of its invention, Bitcoins had lost most of their value, a victim of its own success. With mainstream commercial interest came speculation on the real money markets and then the hacked theft of Bitcoins with a nominal value of half a million dollars. The digital currency dream was over, again.

But there is something compelling about the desire to create a currency that sidesteps all the craziness and corruption of modern finance, one that is genuinely derived from the digital platform. You see it in the gaming currencies and even in the Brixton or Bristol pound. The search for Bitcoin's successor is on.

48

THE RETURN TO CRAFT

The World Wide Web celebrates its twenty-second birthday in 2012. The people who designed the applications that have changed our world are now middle-aged or older and are demonstrating a distinct trend away from digital and towards a maker culture that celebrates artisan skills and crafts. Is this a sign that screen-based creativity is ultimately not satisfying? I don't think it's that simple. From personal and anecdotal experience, I suspect it has more to do with them growing tired of the fact that the craft of programming is invisible to everyone apart from other programmers. Many of the people who built the digital revolution but whose work has never attained the recognition afforded a few notable greats such as Tim Berners-Lee, would simply like to make something that can be appreciated by everyone, their parents as well as their children. I also believe that we are constantly evolving in our ability to interact with the digital platform, and that combining screen-based and traditional forms of work (particularly creative work) is only going to get more and more appealing as we mature in our use of the new digital technologies.

And if there is some digital fatigue amongst these pioneers, it has to be set amid the wider resurgence of crafts such as knitting, sewing, jewelry-making and woodworking, which have been reinventing themselves at least since the late-1990s Stitch-and-Bitch crowd got hold of them. What was once the pursuit of grannies and hippies has gone mainstream cool.

The return to craft is fed by a dose of nostalgia that's typical of an anxious and increasingly austere time, a growing disgust with hyper-consumerism and a consequent fetishizing of the one-off and the homemade. The truly fascinating thing though is the extent to which craft itself has been transformed by the digital revolution. In the Seventies, "homemade" was synonymous with slightly rubbish; now that expertise and professional-standard tools have been made available to the masses, homemade can mean very slick indeed, but still with that appealing originality and quirkiness.

Disenchanted web pioneers bring a particular ability to utilise digital tools in the service of craft, but they are not alone. There's a growing class of people who, tired of the instability of corporate working life in the twenty-first century, are swapping the law for carpentry, or media sales for making cupcakes. They are all exploiting the fact that with digital technology you can now achieve the kind of quality that used to require a factory line. Small-scale production is a return to the idea of the atelier, in which a craftsman or artist works alone or with a small group to produce work of the finest quality. There's a sophistication about the methods and the equipment available now that simply didn't exist before the endless DIY tutorials and master classes available on YouTube, and digital printing presses, 3-D printers, digital knitting machines and a million other tools came within reach of the new generation of craft practitioners.

Then there's the expanding market for these new crafts. The

trend for accessible design has evolved a long way since Terence Conran established the first Habitat shop in 1964. Consumers are more design conscious than ever before, and are coming back full circle, having passed through the age of good quality mass-produced homeware and fashion, via disposable versions, to an emphasis on the value of one-off quality products. This value is partly derived from the increasing popularity of design and craft: people's appreciation of skilled craftsmanship grows as they acquire some practical experience of trying to carve their own table or knit their own jumper.

As the consumers have grown more sophisticated, so has the marketplace for all the things being produced by the legions of passionate amateurs. Etsy, the online craft emporium, is now an e-commerce phenomenon. It allows sellers—90% of them women, the majority of whom are college educated and in their late twenties and early thirties—to sell direct to the customer. There is a staggering quantity and variety of homemade and vintage goods for sale on Etsy. It allows a hat maker in California or a metal worker in Switzerland to reach a global consumer base and trade all over the world.

The appeal of Etsy, and the resurgence of craft in general, is all about challenging hyper-consumerism. It positions itself as resistant to the homogenization of high streets and the impersonality of identikit stores upselling mass-produced, possibly sweatshopped, tat. It is anti-corporate but also extremely entrepreneurial. Etsy is of course a for-profit company, and its sellers illustrate the extent to which the goal of making enough money to live on by doing something you love has become a mainstream aspiration, enabled by the new net-based entrepreneurship.

Etsy sellers might start on a tiny scale—often keeping their day job, or making the most of a home-based revenue stream while they care for young children—but many go on to build highly successful small businesses. Etsy explicitly functions not just as a shop front but also as a business support network that encourages sellers to focus on business as well as craft skills and to consider forming crafting cooperatives when demand outstrips supply. The strength of the model is that a great deal of growth can be accommodated without sacrificing the ideal of buying directly from the person who made the object. There are parallels with the local currencies that encourage shopping locally, but Etsy pushes on a stage further, or perhaps several stages back, to a pre-Industrial Revolution marketplace in which exchange took place between producer and purchaser without intermediaries. Once again, digital technology is being used to re-imagine an old-fashioned ideal and make it modern. It's both progressive and nostalgic at the same time.

The return to craft is partly about shopping differently, partly about working differently, but also about placing creativity at the center of our lives. In the end, perhaps that is its most radical feature. The democratization of creativity is fundamental to the shift to the digital platform. Whether you're blogging your fan fiction or selling your quilts on Etsy, generating your own creative output fosters resistance to the more soulless dynamics of modern life. Etsy won't craft a revolution but, like so many other digital entities, it creates a patch of space in which to be, and all those little patches, networked together, feel powerful indeed.

49

THE INTERNET OF THINGS

Our current Internet is based on a person-to-person model. There may be machines in the middle, but the communication is from me to you via Skype, or from you to your clients via your company website. For years, digital technologists have been imagining an Internet of Things in which machines could talk directly to one another, bringing us into the conversation only when there was something potentially useful or interesting to hear. In the words of Matt Jones, a London-based designer, we're talking about machines that are about "as smart as a puppy"—a concept we'll return to in the Fractional Intelligence chapter.

For instance, imagine a washing machine that emailed its manufacturer, copying you in, when its filter needed replacing. That could trigger a phone call to arrange a time for a maintenance technician to come round and change the part. Your kitchen need never be flooded again. And with Internet-connected software, your white goods could receive updates in the same way as your laptop already does. It used to be the case that the embedded chip that controls your washing machine was decaying towards obsolescence along with the mechanical parts, but with updates downloaded from the Internet, the appliance's software could be repeatedly refreshed, courtesy of iterative design. The life-cycle of your domestic appliances could be extended without any effort on your part.

It's not just washing machines either. The Smart City is dependent on an Internet of Things, with sensors measuring air pollution, traffic flow and the frequency of tube trains, and optimizing responses accordingly. Even goods in shops, (typically clothes) fitted with an RFID security tag are, at least until they are bought and the tag is removed, part of the Internet of Things. And the Internet of Things is itself a precursor to a world in which every object has the potential to be a self-monitoring, autonomously communicating object: in other words, a Spime.

Not every object needs to be connected to the Internet, of course. But design fiction (like science fiction, but making use of technologies already in existence) imagines a scenario in which even your coffee pot might have its own address on the Web. That way, you could connect your next-door neighbor's iPhone to your coffee pot's website so that she could receive a signal that a fresh pot had just hit the right temperature and she was welcome to pop round for a cup. Or your pot could be in communication with your Quantified-Self monitor and send the weekly tally of caffeine intake direct to your doctor's records system.

This might sound far-fetched, possibly not worth the bother, but everything about the advance of digital technology tells us that once something is possible, a useful or entertaining application will be devised to enable you to do it. And once the data that is captured through new applications becomes available, that in turn will be used to analyze behaviors in new ways and drive more technological initiatives.

If you still can't imagine a coffee pot with its own email address, consider the fact that the Lovell Telescope and Tower Bridge are already on Twitter. Inanimate but well-loved objects and landmarks such as these are followed by thousands of people, and they are true participants in the Internet of Things—as long as their steam is generated automatically by bots. (That all changes the moment a human being decides to tweet on behalf of the telescope, at which point you get fewer tweets that say "Obs: B1829-08 18:34:00—08:00:00—pulsar" and more that say "Did you know that @ProfBrianCox once recorded a music video in me?")

In addition to the fact that there are many classes of object that, in all likelihood, we will decide do not lend themselves to the Internet of Things, there is a technological limitation on connection. Or at least, there is at the moment. Every person (or thing) that has an Internet connection requires an IP address, which is their point of access to the vast structure of the Internet. That address is a string of digits. Under the current system, IPV4, each address is a twelve-digit sequence arranged in four blocks of three. That yields approximately 4.3 billion numbers. The problem is that the world is running out of IP addresses. Internet use is expanding so rapidly that a new system, IPV6, is currently being prepared to cater for this exponential growth. IPV6 addresses will use eight blocks of four digits to give a thirty-two-digit number for each address. That's like gaining an extra ten zeros' worth of permutations, and the sum total of all numbers available is so large as to be, for this particular practical purpose, infinite. As usual, the technology is outstripping our capacity to imagine uses for it, but what's certain is that before long, we will have the capacity to network every person and every object on the planet.

50

FRACTIONAL AI

It feels counter-intuitive, but sometimes, a smaller less powerful machine has more uses than a huge one. Take the giant electric motors that powered factories and cotton mills throughout the nineteenth century. Of course they were immensely useful, but only in certain contexts. The potential of the electric motor to revolutionise every area of life was immediately apparent, and there were attempts to use gear systems to drive smaller devices. But it wasn't until fractional horsepower—much smaller motors—was developed that the electric motor became viable in domestic settings, and millions of housewives were released from domestic drudgery.

The same principle that drove engineers to finesse less powerful, more useful fractional horsepower is now being applied to artificial intelligence (AI). Small doses of AI are increasingly embedded in domestic appliances such as televisions, refrigerators and even thermostats to create devices that are, to return to designer Matt Jones' phrase, just "as smart as a puppy."

The original dreams of artificial intelligence designers were on a far more science-fiction scale. The first wave of research, conducted from the 1940s onwards, attempted nothing less than to answer the question, "Can machines think?" Alan Turing's famous test requires a subject to communicate using a keyboard and monitor with both another human being and a computer. If the subject cannot consistently identify which is the person and which is the computer, the machine is said to have passed the Turing test. The ultimate goal was to replicate the human brain, something that remains a very distant prospect.

Subsequently, research in artificial intelligence focussed on manipulating big data. Where the scale of a number problem is beyond the reach of the human mind—in, for example, the super-complex systems of financial markets since the 1970s—algorithms to derive insights that appear to be intelligent have proved immensely powerful. All sorts of features of modern life, such as LoveFilm's ability to recommend titles we might enjoy watching, or credit card fraud detection programmes, make use of this sort of artificial-intelligence application.

But in many ways they are the equivalent of the nineteenth century's gigantic electric motors: vastly complex and expensive. Fractional artificial intelligence works on the basis that most machines don't need to be C-3PO in order to be useful to their owners. A thermostat that can learn your energy-use patterns, for example, will optimize your heating or air-conditioning settings, save you money and conserve energy. Naturally, such a thermostat comes with its own iPhone app that enables you to control it remotely if you decide to stay out for a few drinks after work one night. And if you stay out every Thursday night for three weeks in a row, the thermostat concludes that you are always in the pub on Thursdays and learns to turn the temperature down without you having to think to do it. Smart, but not that smart.

Some of these fractional AI technologies are currently too expensive to be mainstream products, but with Moore's Law in full swing they won't be for much longer. Over the next couple of years we will see tiny amounts of AI appearing in all sorts of gadgets, from the fridge that can complete your grocery order via Tesco's website, to an alarm clock that wakes you ten minutes early if there are problems on the Victoria line. Or fourteen minutes early if it can see that your sleep cycle requires it. Once you have programmed these machines with a minimal amount of set-up information (in the case of the fridge, a typical shopping list; in the case of the alarm clock, your home and office locations) they use fractional AI to adapt and optimize.

Some of the most effective use of fractional AI takes place when it's combined with the mega-powerful data-processing that occurs in the Cloud. For example, Siri the digital assistant on iPhone can rearrange meetings, send emails on your behalf, check directions and, crucially, talk to you in natural language as it does so. Asking "Will I need warm clothes tomorrow?" instantly brings back tomorrow's weather forecast for London, where I am right now. So unlike previous voice-recognition systems, I do not need to learn to "talk Siri": "Access weather; London; tomorrow."

The fractional AI within the device itself is combined with all the power of the Internet, up there in the Cloud, where much of the voice recognition is done. But where the actual computation work is being done doesn't matter, at least now we're all so used to the miracle of the Web. You can't discuss Kant's epistemology with Siri, but the two of you can talk about life's basics as if the machine were indeed able to think.

It even has a sense of humour. Sort of. Should you be tempted to compare it to Hal from the film *2001*, and demand, for example, "Open the pod bay doors, Siri," your long-suffering digital assistant will sigh loudly, coolly indifferent to your wind-up. Try again and you might get a more sarcastic rejoinder. These responses (hidden gags known to techies as "Easter eggs") have been coded by the clever programmers who anticipated your jokes long before they occurred to you. They are proof of human beings' ingenuity, not the machine's. And yet, since Siri can learn, it might not be long before it's genuinely answering back. Hal is still a sci-fi nightmare, but Siri is already quite a bit smarter than a puppy.

51

WAR ROBOTS

Of course, some situations call for something a little tougher than a puppy-like PA. The Armed Forces have been using drones (flying robots) for years to carry out reconnaissance operations in hostile territory such as Afghanistan. It is far cheaper, and of course less risky, to fly a remote-controlled plane with a camera strapped to it over enemy terrain than to send in a manned aircraft. The original drones typically had a wingspan of about four feet and were controlled by soldiers in the field. As the technology grows more sophisticated, they are more often piloted by personnel back at base.

The U.S. military is investing particularly heavily in drones. From 2010 it trained more drone pilots than fighter pilots. A drone's control mechanism is very similar to the games consoles that most of the young pilots have grown up playing. A nineteen-year-old with thousands of hours clocked up on his PlayStation already has all the manual dexterity and hand–eye coordination needed to pilot a drone.

Increasingly, this new generation of drones is used not merely in reconnaissance, but in combat. The long-endurance multi-intelligence airships that the U.S. has recently begun to commission are not just spy blimps. They carry weaponry as well as cameras and can stay in the air for up to three weeks, following pre-programmed routes determined by GPS systems and controlled by pilots who might easily be on the other side of the world. They are particularly useful in fights against insurgents, where the enemy has little access to ground-to-air missiles.

But the resulting depersonalization of war and the asymmetry in the terms of engagement raise some troubling ethical considerations. Combat drones have been implicated in several recent assassination attacks against Al Qaeda targets. Is it right to attack an enemy using a device that cannot itself be killed?

Traditional military technology is extremely expensive and a closely guarded secret. It tends not to "water down" to applications for civil authorities. But drones are simple machines and extremely cheap. The drones available on the open market range from £1,000 for a top of the range model to £300 for something that gets the job done. Which makes them very attractive indeed to domestic law-enforcement agencies.

In 2011, London's Metropolitan Police Force applied for permission to the Civil Aviation Authority to fly drones over the crowds at the 2012 London Olympics. Where security concerns are crucial, it is easy to understand the rationale of using drones, but as we have discussed elsewhere, 2012 will also be a year of protests. And the temptation to use drones to survey protesters who have not in fact committed any crime will likely prove irresistible. Given the extent to which the police already use infiltration techniques, not to mention manned helicopters, the use of drones is hardly a radical departure in terms of technique. But there is something potent about the symbolism of a Big Brother eye in the sky.

52

CYBER WARFARE

In the twenty-first century, the Internet is the fifth battlefield. Since much of the machinery of nation states has shifted online, it follows that states will defend and attack there as well as on land, at sea, in space and in the air. But cyberspace is full of other entities—corporations, hacktivist collectives—just as capable as a nation state of participating in cyber conflict, all of which have contradictory interests that they are prepared to defend. This complicates the dynamics of cyber warfare considerably.

It is not of course a novel development for conflict to be carried on in arenas that draw civilian agents into play. The difference is in the new balance of power. Military might is still mighty in cyberspace, the nation state's defence budget being what it is. But the levelling effect of operating online (where, as we saw when looking at twenty-first-century statecraft, you are only as powerful as your programmers, and all websites are pretty much alike) means that an army doesn't dominate as it would in a traditional physical setting. There are new battles in cyberspace, pitting nation states against corporations, corporations against hackers, and hackers against everyone else.

Cyber warfare is carried out by nations both on its own and alongside conventional warfare. The jostling for military supremacy in cyberspace has been going on for years now, to the extent that there is constant probing of Western governments' online defences by countries like China, and vice versa.

During the brief Russo–Georgian war of August 2008, the Russians used cyber warfare to temporarily bring down the Georgians' networks as part of their strategy of full-spectrum targeting. They had learned their lessons from observing the 2003 invasion of Iraq, when vast sections of infrastructure on which the enemy depended were destroyed by the coalition forces. In the aftermath, the coalition found their mission hampered by the loss of these crucial communications networks. Hacking a region's network to close it off, rather than bombing it, is (temporarily) highly effective, and unlike a bombing campaign it's reversible. But there are also clear limitations—not least the fact that this technique of cyber warfare is exactly the same as that used by oppressive governments to deny their citizens' human rights.

There has been a scramble to invest in cyber warfare technology. The appeal is compelling, and the short-term gains can be enormous. But at root, cyber warfare is just hacking, and the thing about hacking is that the network is designed to route traffic around the site of disturbance. The only way to operate effectively in cyber warfare is in situations such as the Russians found themselves, where they were prepared to take down a whole country's network. Its use is problematic, and counter-measures are no easier. There are now numerous cyber threats to the national defence in countries such as the U.K. and the USA that, were they responded to in an effective way, would negate the values those states were trying to defend.

As the nation states struggle with technical and ethical questions, other entities exploit the growing expertise and the cheaper software and begin to behave in ways that would previously have been associated only with national governments. Anonymous, the loose collective of hackers and activists we met earlier in the book, brought down the Libyan government's computer network during the recent civil war. Even five years ago, it would have been unthinkable for a citizens' collective on the other side of the world to attack a nation state in this fashion, let alone to succeed.

The French state has been shown to be targeting foreign-owned corporations in competition with French corporations. There has been some suggestion in Britain that the Foreign Office should adopt a similar strategy. What used to be known as industrial espionage has on occasion escalated into full-blown corporation-on-corporation cyber warfare.

Cyber warfare requires us to think in different ways about how combatants engage each other. If it has any equivalent in the flesh-and-blood world, it is perhaps the (counter-) insurgency. The appropriate metaphors are epidemiological, or derive from fashion. In cyber warfare you seek to infiltrate and influence and inspire. Your targets move like viruses or trends, not like battalions. It's all about the prevention of the spread of data. One of the most difficult-to-grasp concepts for large-scale entities, especially nation states, is that, in keeping with the non-hierarchical nature of the medium, in many ways the defence against online war is the same as the defence against spam advertising. It has to be conducted at the level of the individual user, who must be equipped with their own counter measures. This, again, pulls the military closer to other online interests such as brands. Twenty-first-century networked power means that in cyberspace at least, two hierarchies will never again confront each other in a bilateral encounter.

In fact, as we'll see in the next chapter, the biggest threat could come not from an enemy state, or even a terrorist organization, but something even more sinister . . .

53

THE SINGULARITY

At some point, human beings might create an artificial intelligence that is smart enough to design its smarter descendent. Within a few iterations, that intelligence would have reached superhuman levels. This AI would not only have extremely sophisticated learning abilities (as humans already do) it would by then be running on a physical platform infinitely superior to the hardware we humans carry around in our heads in the form of brains. The inevitable exponential growth in its capacities would leave us infinitely far behind and its progress towards total power would be assured.

All sorts of sinister outcomes suggest themselves. An omnipotent being might treat humans as creatures to be hunted, or farmed for food, or used as slave labour. There is nothing to say that the being would be hostile, of course, but in any case we have far outstripped our powers of speculation. Trying to imagine what might occur if we encountered this being is like asking an ant to imagine listening to Mozart.

That's the Singularity, a term coined in 1993 by Professor Vernor Vinge of San Diego State University, computer scientist and acclaimed science-fiction writer. Vinge thought it would happen not before 2005 and not later than 2030.

The Singularity reeks of the sort of thing teenage me dreamt up at 4 a.m. after a big night out, but Vinge's ideas are not merely the stuff of science fiction. When he published his paper on the Singularity he stressed that superhuman AI was only one of four ways that a Singularity might occur. The other three, which he was clearly less excited by, are nonetheless just as alarming in their implications for our welfare, and, to most people's way of thinking, also far more plausible. The second scenario is of an accidental Singularity, in which a large networked computer system (and its associated users) may stumble into Singularity. Then there was the possibility that the interface between individual users and their computers becomes so porous that the users could themselves be considered superhumanly intelligent. Or, last of all, if biological science found effective ways to improve human intelligence to superhuman levels without any involvement from computers, then the conditions for a Singularity might still be met.

The Terminator remains an unlikely threat, but that shouldn't leave us feeling complacent. Just because the super-intelligent being doesn't look imminent, doesn't mean a super-complex one isn't already capable of harming us. We have started to see the kind of potential for catastrophe brought on by a prototype accidental Singularity, and it's not at all pretty. A large enough collection of algorithms interacting over a big enough network can produce some very peculiar and painful results indeed, as the traders on the Chicago Stock Exchange discovered first-hand during the Flash Crash of 2010. Readers of the chapter on High-Frequency Trading will already have figured out that what happened on that day in May when the Dow Jones index lost 600 points in three minutes was, precisely, an instance of a pseudo-Singularity at work. When the algorithms and their interactions have got so complex that it takes a team of experts five months to figure out what happened during a five-minute event, we are in trouble.

Then there's the What If factor attached to a huge growth in fractional AI. A world in which you have a billion machines all as smart as a puppy could easily feel as chaotic as living with a billion puppies. We have very little idea how any of this AI interacts with itself, because it's all too new and, at the worrying end of the AI spectrum, too complex.

What we do know is that we are already creating a new kind of weather, composed of weird feedback in loops between Algos. The Internet is its own landscape now, and tiny events can have significant consequences. It's possible to imagine a chain of events that goes like this: the London train network is disrupted, so everyone in Finsbury Park with an Internet-connected alarm clock gets woken at the same time to head to catch a particular train. When they arrive at the station, the train has been cancelled so they converge on the bus stops where a pavement density sensor is triggered. Extra police officers are drafted in from Camden to Finsbury Park. Crime spikes in Camden as a consequence, the stats are automatically published online, which leads to a drop in property prices and a rise in Camden council's bond prices which means they can't afford to open the new hospital ward they had planned and people die for lack of treatment.

This is obviously only a hypothetical example, simplified for ease of understanding and exaggerated for effect. But the core problem is real enough: in addition to all the randomness of life that humans generate themselves, we now have to contend with a potentially chaotic complexity orchestrated by artificial intelligence interacting with other agents.

There are several points in this scenario where human beings might well have diffused the situation: if enough of the commuters waited on the platform or headed for the Underground as well as the buses, for example, overcrowding wouldn't have become a problem. Or, if it had, police officers might have managed the crowd swiftly and returned to their Camden posts long before the local burglars had time to commit enough crimes to change the stats significantly. But the fact that this particular scenario might well not play out does not mean that it couldn't, and if it did, given that it would be infinitely more complex than we can follow on a page, it would be impossible to trace the deaths in Camden to the alarm clocks in Finsbury Park.

Millions of other feedback loops are occurring at this precise moment online, and as the Internet accumulates more functions, more Spimes, more AI and, crucially, more data, their potential to cause disruption is growing exponentially. There will be many more incidents like the Flash Crash. The fact is, the Singularity is already with us.

54

NET NEUTRALITY

The Internet has no central authority or presiding genius calling the shots or making decisions—it is by its very nature a rough consensus cobbled together through trial and error. Technical standards have emerged out of working and reworking the code, not the other way around; but the whole enterprise has nonetheless been built on certain core principles. One of those is that all traffic across the network is equal. Data that will end up as an email has the same priority as data that will end up as a video stream. Your email has no more priority then my email, and is treated just the same as President Obama's.

Net neutrality stems partly from egalitarianism and mostly from fundamental tenets of network design. The end-to-end principle dictates that while it's travelling, data is just data and its content is invisible to the network it passes across. That data could be an email to your boss or a video clip of a kitten doing something cute or a confidential memo from the President's office. But once it has been broken down into packets of data, that email or that video has no outward markers of the application it was made with or where it came from. Anything application-specific happens only at either the disassembly or reassembly point of the network.

This basic truth has dictated the way in which the Internet develops and is used; it is the reason why, as we have already seen, it is impossible to censor the Internet without effectively breaking it. Many of the Web's most thrilling or disturbing characteristics, depending on your point of view, derive from this principle of neutrality. But it suits end users and content-generating companies far more than it does the Internet service providers (ISPs) whose business derives from the neutral middle. Increasingly unhappy, they are calling for a system of prioritization so that certain content can be regarded as premium, and priced accordingly at the point of transmission. This poses significant risks for both users and content companies, but the ISPs believe they have a compelling argument.

Telecoms corporations invested heavily to develop and install the fiber-optic cables that brought us all the miracle of high-speed access. As this technical capacity has grown, so too has the number of users, their appetite for music and video-based content and therefore the volume of digital traffic. Demand for the Internet service providers' product is constantly outstripping supply, but there are frustratingly few mechanisms, from the ISPs point of view, for making money out of that fact, and simultaneously it places them under permanent pressure to spend on bigger and better infrastructure. Users luxuriate in services that are dependent on constant high-speed Internet connectivity, such as video Skype or BBC iPlayer, and it's those companies—and others such as Google, Vimeo or Facebook—that derive the lion's share of the profit.

The ISPs insist that the big content providers should be picking up some of the tab for the outlay on network infrastructure, and are lobbying to be able to charge them a premium to send their huge volumes of data over the network via high-speed connections. After all, a significant proportion of all U.K. Internet traffic after 7 p.m. derives from BBC iPlayer. So far the content companies have successfully resisted by making reference to the net neutrality principle: if high-speed access has to be paid for beyond a certain volume of traffic, who is to define that limit? It's easy to imagine a scenario in which most non-corporate entities are unable to post video or music files to the Web, a situation intolerable to end users.

Charging would also necessarily tend to produce monopolies and stifle the entrepreneurship and creativity that has characterized the Net's development until now. If you need to have Google-sized purchasing power to afford the necessary access to high-speed connections to develop your company, your innovative new search-engine start-up is highly unlikely ever to reach critical mass.

The alternative—to charge end users different rates for different packages of Internet connectivity—quickly runs into similar problems. Now that the Web is the dominant platform for all sorts of activities that liberal societies consider sacrosanct, such as freedom of expression and freedom to gather, the prospect of some sectors of some societies being cut off from the Internet is, again, intolerable. In an age where Internet access is a de facto human right, for ISPs to segment their market, a standard business technique in other sectors, is highly problematic.

Net neutrality is clinging on but the Internet service providers have not given up the fight. The discussion is ongoing and is now shifting into the political arena. We cannot afford to treat the matter as a spat over a business model. The implications for the way in which the Internet develops in the twenty-first century are too profound for that.

55

PUBLIC-KEY CRYPTOGRAPHY

There's nothing new about public-key crypto. It's been around for as long as people have needed to send secure messages across the Internet: to their bank, to a retail site, for confidential business dealings or military planning. But aside from being a fundamental aspect of how the Internet works, and an elegant idea worth appreciating for its own sake, public-key crypto also tells us something counter-intuitive but fundamentally true about security in the digital age.

We instinctively feel that as more aspects of our lives shift online, there is more scope than ever before for confidential data to leak. If our first thought is that we must work harder to keep things secret, we're not yet attuned to the new reality. Secrecy is only one component of modern-day security, and it's not the most important. These days, real strength is obtained through sharing, not obscuring, your security tools.

Almost all traffic over the Internet is now encrypted, to protect financially sensitive information and privacy, and guard against eavesdropping. The mechanism of this encryption and decryption is designed so that only the sender and the intended recipient can use it. This is achieved through deployment of a beguilingly ingenious concept: public-key crypto.

Before the invention of public-key crypto, two individuals needed to have a prior relationship that would allow them to agree a code so that they could send an encrypted message. Given the volume of business now conducted over the Internet, we need access to a code-generating system that gives us confidence that our message will not be intelligible to a third party and that does not require us to be constantly contacting recipients to agree an encryption method before we send it.

That system works like this: let's assume that Alice wants to send Bob a secure message. Both have two keys—one public, one private—which are unique to themselves, as well as a password. (A key is in fact a very large number, composed of strings of digits, and generated by an algorithm.) Alice can take Bob's public key, run it and her message through an algorithm that spits out the encrypted text, and send that text to Bob over the Internet. Bob then has to use his private key, along with his password, to unlock the text, decrypt it and read the message. When he wants to reply to Alice, he does exactly the same thing, using her publicly accessible key, and Alice reads the message by decrypting it with her private key. So long as Alice and Bob keep their passwords secret, their security is assured and they can send anything to anyone in the world. A vast field of secure communication is opened up, with all the potential that implies.

Hypothetically speaking, there are two ways that this system could be broken. The first is to find a mathematical flaw in the algorithm used to encrypt and decrypt the message. The second is essentially to apply brute force, by working through every single possible private key. To give a sense of why none of us need to be worrying about that second possibility, it's necessary to appreciate the size of the numbers involved. A private key is measured in bits and a standard size is 256 bits. The number of digits in a private key is 2 to the power of n where n is the number of bits. A standard private key is one of 2 to the power of 256 permutations. If you had a computer that could check one billion billion keys a second (and a computer this powerful does not exist) it would still take 30, 000, 000, 000, 000, 000, 000, 000, 000, 000, 000, 000, 000, 000, 000, 000, 000, 000, 000, 000 years to try every single possible number, by which time, one assumes, your credit card would have long since expired.

That's the reassuring part. It also means that all the strength of public-key encryption derives from the particular algorithm at its heart. Before a new algorithm can be used, it must be rigorously tested, and this means that basically it is put out to the encryption community with an invitation to break it. These algorithms for generating private keys are deliberately made very public so that the weaker ones can be weeded out, which feels counter-intuitive until you remember that the need for privacy derives from the number the algorithm generates, not the algorithm itself. Sometimes a competition is held and algorithm designers' submissions are whittled down over a number of years until a winner emerges that no one can break.

Public-key encryption depends for its formidable effectiveness on a combination of a relatively small amount of secrecy, combined with a vast amount of openness. This is a given in the digital world—that one's security should not depend solely on one's ability to keep certain pieces of information hidden. In an age of more and more publicly accessible data and ever more skilled hackers, security through obscurity is obsolete. Experience has shown that in fact precisely the opposite is true: your security depends principally on the cleverness of your algorithms, and that can only be demonstrated by public scrutiny.

It's standard practice for flaws in operating systems, for example, to be broadcast as widely as possible. If you are a digital security worker and discover that Windows has a gaping hole in its operating system, once you have informed Microsoft it is considered to be your moral duty to tell the world at large, as fast as possible. Yes, that might leave millions of users exposed to security breaches, but the fact is that if you have discovered the flaw means that in all likelihood so has someone with intent to cause mischief. News of such security breaches travels very fast indeed on the Internet and it is far better to inform users so they can take preventative measures, or respond quickly to damage.

The efficacy of online security is yet another derivation of the Internet's truth zero: power is out of our hands and dispersed across a vast network, but that doesn't mean that we ourselves are totally vulnerable. If you want to operate safely on the Internet, paradoxically, you have to trust that diffuse power, not resist it. Understanding its ways will enable you to make rational decisions about how to protect what is most valuable to you. Attempts to erect an old-fashioned fortress, on the other hand, are doomed to failure.

56

THE DARK NET

The World Wide Web is far, far larger than you think it is. It's easy to assume that because search engines like Google can reveal virtually anything we want to know or could possibly imagine within 0.3 seconds, we can see everything there is to see online. In fact, perhaps as much as 90% of material that's posted to the Web is invisible to standard search engines. (Although, who knows since it's invisible—it could be even more, or much less.) One thing's for sure: there are dark places all over the Internet, places you can only find if you know how to look for them with special tools and security codes. And in some of these dark places there are very strange and very nasty things lurking.

It's not all a Gothic nightmare, mind. As always with our digital world, there are an infinite variety of ways to use the technological capabilities and, in this instance, many degrees of seclusion. Some of the dark places are havens of peace rather than subterranean caverns. Many of the archives of academic journals, for example, are online but have not been indexed by Internet search engines. Their material is accessible to specialists who know their location but is not connected to the hullabaloo of the everyday Net. Sometimes a tiny step off the neon-lit super highway is enough to access a calmer world. Google Scholar, for example, which does not feature anywhere on Google's home page, allows anybody to search across a vast catalogue of PhD theses, court records, professional societies' papers and academic publications. All over the Internet there are flourishing communities and repositories of material that do not show up on conventional search engines. Many of them are benign, analogous to private members' clubs. But not all of them.

Over the last ten years the sophistication of communications apps has been increasing exponentially, and one of their key features is a trend towards a very high level of anonymity. Many of these apps are developed by security agencies to resist espionage: to protect users' identities and prevent eavesdropping online. They can be deployed to protect national interests and promote freedom of speech in sensitive regions but they can of course also be used by agencies and people who wish to evade the attentions of what we might be tempted to call the good guys: accountable authorities such as the police in democratic regimes, watchdogs of all kinds. As we've said before, there are many legitimate reasons for people to wish to be anonymous online, but there are also plenty of uses of anonymity that might be considered suspect, depending on your political stance. And then there are the out-and-out criminal reasons to be anonymous, notably the trade in hard drugs and child pornography.

There is a long-standing argument about giving the public access to high levels of anonymity. There is no doubt that bad things happen on the Dark Net, as they do in closed spaces all over the flesh-and-blood world, but our response to that needs to be reasoned, not panicky. The worry that terrorist cells will use the Dark Net to plan their latest atrocities is an understandable one, but it doesn't stand up to much scrutiny. The fact is, there are many ways to pass secret messages. The communication apps that are freely available to all of us at the click of a Google-powered button are more than up to the job. Flickr photo accounts are used to pass encoded messages via steganography, in which numbers sequences are embedded in the digital code of photos. Or, more simply, a narrative sequence of red cars and certain zoo animals can be agreed upon and posted to Flickr to signal the time and location of an incident. Twitter is a gift for message passing. And it's fun to come up with your own techniques. The problem is the same old one of Internet censorship: if you want to prevent some people from misusing the technology, the only way is to close it down to everyone. Or to do good police work and consider the communications apps your tools to trap the criminals.

That there are many criminals operating online is beyond dispute. Occasionally their activities are exposed to the light and we catch a glimpse into the abyss. In 2011 for example, a retail site on the Dark Net became notorious after leaks of information to the conventional Web caused an outcry. Silk Road was described by U.S. senators as "the most brazen attempt to peddle drugs online we have ever seen."

Silk Road is a marketplace site allowing individual sellers to offer contraband goods, particularly drugs like heroin and LSD, in exchange for Bitcoins, the anonymous digital currency we encountered in an earlier chapter. The site is only accessible via a piece of software called Tor which, ironically, was developed and is maintained with funding from the U.S. State Department to facilitate eavesdropping-proof, ID-anonymous online communication. Tor is freely available as a download, and once installed sites can be hosted and accessed anonymously. It works by sending all information around the Tor network before passing it out to the conventional Net via a randomly selected terminal, obscuring the connections between individual users.

When news of Silk Road leaked, site traffic massively increased and the value of Bitcoins shot up. The site's operators, whose policy prohibits the sale of goods or services intended to harm others, reacted robustly to calls for their closure. They vowed to carry on and, having employed all the most cutting-edge technologies for remaining anonymous, at the time of writing are still trading—though for how much longer seems uncertain.

Other Dark Net activities are so toxic that their existence remains far better hidden, but once exposed the perpetrators are certain to be pursued. There are complex police operations in progress all over the world that use the Internet as their most powerful weapon in the hunt for those who make and consume child pornography. The Internet, and increasingly the Dark Net, are here both the mechanism for delivery and also the channel for detective work.

Rigorous policing can yield impressive results, but so, it seems, can clever hacktivism. In October 2011, our old acquaintances the Anonymous collective found something that greatly displeased them. The site was called Lolita City; it was the largest single site selling child pornography ever uncovered and it could only be accessed via an intermediary website called Hidden Wiki, which in turn could only be accessed via the Tor network. When Anonymous found it, they contacted the web host, Freedom Hosting, to demand its immediate removal. When that was refused, Anonymous hacked the site. The web host restored the content within twenty-four hours and added new security details. So Anonymous hacked them again, and this time they published the user names and credit card details of 1,589 pedophiles. They have declared Freedom Hosting "OpDarkNet Enemy Number One" and sworn to continue to crash their server and any other they find that contains, promotes or supports child pornography. Online vigilantism has its detractors, but on this occasion most people would feel an instinctive sympathy for the cause. And as usual, there is something thrilling about Anonymous' capacity to get things done.

The Dark Net has some truly vile corners but new cyber life forms are evolving, capable of fighting back. The result is all manner of strangeness. Before the story broke, many people could never have imagined that a bunch of cyber-anarchists would take down a pedophile ring on a secret Internet using software funded by the U.S. military. But then again, perhaps we shouldn't be surprised: the Internet is like that.

57

WHY MONITORING ONLINE MESSAGES IS NOT THE SAME AS WIRETAPPING

Politicians in government must have the most anxious relationship with Internet technology of any class of people in the world. Parents of pre-teens might come a close second, but it's our elected representatives that are seemingly the most troubled by the threat to authority it poses, and the most desperate to control and monitor other people's Internet use. Not that I'm implying that parents are wrong to be concerned. It is of course perfectly understandable that you might worry about who your twelve-year-old daughter is talking to on Facebook, or in a chat forum. But as most parents quickly work out, the way to keep your children safe online is not to ban chat forums—she'll just access them at school, at a friend's house, in the public library—but through good parenting that allows her to make sensible decisions about her online behavior.

Politicians' anxieties are far less justifiable. I'm writing this in early 2012, and the policy makers in many countries, the U.K. and U.S. especially, are busy proposing laws to extend their state's abilities to monitor Internet usage. One of politicians' claims is always that much more Internet surveillance is needed in order to protect us from serious crime. This is, in my view, completely spurious, and it is worth addressing just why this is.

Let us consider a hypothetical example. The U.K. security services already have the legal and technical ability to know who you are communicating with via the telephone networks. The actual contents of the conversation are subject to much stricter checks and balances, but who you called and when is straightforward for them to discover. Now imagine you wished to have the same information, but for communication on the Internet rather than the phone: to know for sure who an individual was communicating with online. As we move into a pure Internet world, we, and the security services, are beginning to discover a terrible truth. Their capabilities are declining: it is actually impossible.

Sending messages via the Internet is not like picking up a phone and calling someone. The analogy of an subpoenaed itemized phone bill that reveals the duration and recipient of a call does not apply. There are two simple and unassailable reasons for this.

Firstly before the web there were very few ways to send a personal message online. There was email, and a few instant messaging systems, but all sent their messages, not via the Web (which, lest we forget is an application that runs on the Internet, and not a synonym for the Internet, even if we frequently use it that way) but via their own specific set of technologies, meaning that you could identify and separate their data from all the other kinds of online traffic with ease. These days, most interpersonal messages are sent not via specific messaging systems, but via applications that are part of the Web: Facebook, Twitter, or your favorite web-based email provider. These are indistinguishable from all the other traffic on the Web. If you want to be sure that you're collecting all of the interpersonal messages that are travelling across the Internet in your country, you therefore have to record every single bit of data that goes over the network. You then have to have the ability to take this recorded webpage data and understand which part is a message, and which is part of the site itself. That in turn requires a perfect understanding of the page structure of every website where it is possible to send someone a message. This is not practicable.

But even if it was, the second reason that our imagined legislation is impossible even to introduce, let alone enforce, is that virtually all traffic over the Internet, including email, and most web-enabled applications such as Facebook, is encrypted. This means that even if a government did decide it was going to access content in order to identify Facebook messages, their scanners would be at a loss. As we've seen, public-key encryption is a thing of beauty, and very good at its job. There is no, and can be no, system in existence powerful enough to break it even on an ad-hoc basis, let alone in such a fashion that would give security agencies the access they need to carry out their stated aims. Social media sites started to encrypt their users' messages in 2011, in large part due to snooping by the toppling regimes of the Arab world. They were encouraged to do so by liberal democracies, including our own.

Which brings us full circle, via the technological reasons that it can't be done, to the civil liberties argument. What's bad for a dictatorial regime or a fledgling democracy is also bad for a well-established liberal democracy, and we should resist any complacent attempts to delude ourselves otherwise. Why should the emerging democracies in the north of Africa take us seriously when we pass the kind of draconian legislation they have spent years struggling to escape?

It is depressing to see that even fifteen years after its wide-scale adoption, many of the people who run our country still haven't grasped that the World Wide Web is not the same as the Internet. I tell myself that it must be a kind of willed ignorance rather than anything more sinister, but sometimes even I, a staunch non-believer in conspiracy theories, catch myself wondering. Meanwhile, the ill-conceived, technically laughable, financially impossible laws keep on coming.

58

CARBON AND THE DIGITAL ECONOMY

There is a persistent rumor that our Google habit is going to cost us the Earth. The story made a big splash back in 1999 when a piece in *Forbes* magazine claimed that Internet usage was responsible for 8% of the USA's total annual electricity consumption and that figure would grow to 50% within ten to twenty years. The fact that this wasn't true at the time and has been proved spectacularly untrue since does not, unfortunately, mean an end to the matter.

It should not surprise us that there are people whose interests are so threatened by the radical changes brought about by the Internet that they will back the feeding of disinformation, of whatever kind, to the press. And neither should it be surprising that many people have an intuitive suspicion that this sparkly new economy, this driver of growth and changer of lifestyles, must be consuming vast amounts of energy. The IT industry (and more particularly, the Internet) as a source of ecological damage on the same scale as the aviation industry is a persistent, guilt-inducing meme. But it's simply not the case that your web searches are a major contributor to climate change.

Firstly, of course, it is essential to recognise that the IT industry, like any other, has a carbon footprint. So do our personal computing habits. Computers, smartphones and tablets are all electronic devices, which means that it is important to make them (and the data-centers that serve them) as energy efficient as possible. But just because there is a carbon cost attached to the shift to a digital world, does not mean that the cost is vastly larger than that of those industries the Internet is replacing; in fact, just the opposite is true.

To assess the environmental impact of the shift to digital, it is necessary to consider something called energy intensity, which measures the amount of primary energy consumed per dollar of real gross domestic product (GDP). This allows us to compare energy use in the new economy with what predated it. And when we look at the situation in the round, like this, the numbers are startling. The Energy Information Administration, a U.S. government agency, reports that energy intensity dropped by 7.5% between 1986 and 1995, as one would expect of an economy moving from being industry- to services-based. The startling thing is what happened next: a 20% drop in energy intensity over the next ten-year period, between 1996 and 2005, when the U.S. economy changed gear again. These were the years of the build up to the dot.com boom, and the mainstreaming of e-commerce, as the Internet wreaked its changes across countless industries.

Internet-driven growth accelerates an existing trend towards energy-efficient growth that happens as developed societies move away from heavy industry. It depends largely on dematerialization: turning the product of the music industry from CDs to downloads, for example. This has enormous knock-on savings not just on the carbon cost of the primary materials, but on their manufacture, distribution, warehousing and disposal. A 2009 study into the energy consumption of the music industry pre- and post-digital which employed total accounting to assess all these factors has concluded that the (now digital) music industry's energy consumption is at least 40% lower than it was when it was selling CDs. These levels are typical of other e-businesses. Even online grocery shopping is energy efficient. One electric van delivering to thirty households in a single day saves a lot in fuel bills. So, while carbon costs are, in a coal- and oil-dependent world, a corollary of any kind of growth, some forms of growth are less toxic than others.

There has recently been a great deal of anxiety about the negative environmental effects of the shift to Cloud computing. The computing power of a smartphone is enormous, but its physical dimensions are not. Now, as the expression goes, the network is the computer. The computational power of your smartphone resides in the networked data-centers whose energy consumption we are all supposed to feel so bad about. But there's another misunderstanding here. These places run at approximately 90% efficiency, which means that though, obviously, the amount of energy they consume is vast, it's still more efficient than having the equivalent number of home PCs working at perhaps 5% efficiency to get the same amount of computing done. And with heat-capture technology, much of the energy the centers waste can be recycled. They are not exactly models of self-reliance yet, but the IT industry is nonetheless a considerably more responsible carbon consumer than most, especially relative to how much power it pumps into the economy.

In addition to all the high-energy-intensity growth that's been displaced by the Internet, and the ever-increasingly energy-efficient hardware and practices it fosters, there are numerous other features of the digital world that make it a less ecologically toxic place to live. In a Smart City, where there is open access to information, energy use can be monitored and optimized. The Internet of Things means that the built environment is full of objects designed to run on as little energy as possible. Digital metering systems for power and water in office buildings and homes, computer controls in car engines that reduce emissions: certainly in the developed world, the Internet has produced a net reduction in carbon emissions.

One of the biggest plus points of the Internet, in terms of its positive contribution to environmental responsibility, is harder to measure than the energy use of an office block. It lies in the information dissemination it excels at. It's impossible to imagine that there would be a globalised, flourishing, practical, grassroots environmental movement without the Internet. The Transition Towns movement that we will look at later in relation to local economies, for example, are typical of the kind of information-sharing initiatives that translate to real change, in large part thanks to the Internet.

You could argue that despite all this, and despite the financial crises that dimmed our faith in capitalism at the end of the first decade of the new millennium, we are still addicted to consumption and to growth. It is true that the Internet has been the crucial element in the new economy, and that the new economy is guilty as charged of poisoning the planet. But it does not follow that this is all the fault of the Internet. In fact, digital growth is cleaner growth. And we can take some comfort from the fact that the Internet is enabling some progress on one of the biggest blocks on unilateral agreement to binding carbon emissions cuts. Developing countries had a killer argument when they pointed out that it was unreasonable for developed nations to deny them the chance to grow their economies and lift their citizens out of poverty, all in the name of environmentalism. Recent years have demonstrated that digital industries can enable developing nations to leap-frog heavy industry and aim squarely at cleaner, safer, more sustainable growth.

Of course, the Internet is no panacea. Only one of the most obvious problems is that an IT-driven economy requires a highly educated workforce, something that many developing nations lack. But nonetheless, in a complex situation that poses the greatest challenge of our time, IT is not the enemy. As Joe Romm, author of what is still the standard study, *The Internet Economy and Global Warming*, says, "If you are worried about your carbon footprint, buy one hundred per cent green power and do an efficient retrofit on your house . . . and let the Internet keep saving people energy and resources."

There are plenty of reasons to feel anxious about climate change, but the world's Internet use is not one of them. Keep right on Googling.

59

GEOENGINEERING

It is a generally accepted fact, with which I have no argument, that man-made climate change poses the gravest and most complex challenge of our times. Aside from a disproportionately vocal minority, pretty much everyone agrees that it's happening, but nobody can agree on what to do about it. The first of many dilemmas boils down to whether we should attempt to reverse the damage that has already been done and prevent further problems, or focus our efforts on managing the fallout of the inevitable warming.

Thus far, the world has paid lip-service to the idea of reversing and preventing damage by reducing carbon emissions, but with scant concrete progress. The alternative interventions now being investigated are very different from the mass retrofitting of the entire industrial complex. Rather than making industry, buildings and transport cleaner and greener, these techniques would engineer the climate itself by mimicking the dust clouds produced by volcanoes or installing giant space mirrors.

Suspend your disbelief: there is serious research from highly respected institutions underway and though much geoengineering is prohibitively expensive and liable to cause side effects that are as bad as or worse than climate change, some of the more pragmatic proposals could be genuinely helpful. Not that we should indulge in a collective sigh of relief. As Professor John Shepherd, who chaired the Royal Society's 2009 report into the viability of geoengineering put it, "None of the geoengineering techniques is a magic bullet. It is essential that we strive to cut emissions but we must also face the very real possibility that we will fail." In that case "Plan B," geoengineering, might be the only hope.

There are two main ways to achieve Plan B: through carbon dioxide removal (CDR) or solar radiation management (SRM). The former is by far the better option because it addresses the problem rather than merely mitigating the warming that results. Reducing carbon by either converting it into something else or capturing and storing it safely would reverse and prevent damage, and would tackle related issues such as the increasing acidification of the world's oceans. Reflecting more of the sun's energy would lower temperatures but do nothing to fix the underlying issue.

Unfortunately, so far it seems that carbon dioxide removal is way more expensive and challenging from a technological point of view than solar radiation management. Carbon capture at the source of release, i.e. power stations and other major polluters, is considerably more viable than a mechanism to hoover up free-floating carbon in the atmosphere and scrub it clean, but even so, it's estimated that it would increase energy costs to the consumer by up to 90%. Even if you could suck up all the carbon you needed to (and that machinery itself would require vast amounts of energy to run) you would then have to store the captured carbon safely in underground deposits, and hope that there were no disastrous unforeseen consequences. It's safe to say that an awful lot more research and development is required.

Another smaller scale but potentially more practical carbon removal technique involves enhanced weathering. This could be a good long-term strategy, though, again, it's still at an early stage of development. One entertaining strand of research is being carried out by Dr. Rachel Armstrong, who has a medical background but now works at University College London's architecture school. She has created a paint that contains biological material that reacts with carbon dioxide in the atmosphere to produce calcium carbonate—limestone. Potentially, the smart surface of a building could absorb CO_2 and convert it into layer upon layer of its own shell, made up of limestone deposits.

By comparison, solar radiation management is far more practical and affordable, even though its applications can sound a little science fiction—space mirrors, anyone? The problem is, it's also nowhere near as useful and much riskier.

The most credible method for managing the sun's power is to imitate the cooling effect produced by volcanic explosions, in which vast dust clouds are shot into the atmosphere. The idea would be to introduce huge amounts of sulphur dioxide into the upper atmosphere over the polar icecaps, where it would form tiny particles of sulfate that would reflect more of the sun's energy into space. This could produce a cooling effect of one or two degrees, enough to allow a refreezing of the icecaps and to stall the series of tipping points that, once triggered, might lead to cataclysmic change.

This sounds wacky but would in fact be relatively easy to achieve. A University of Calgary study got good results when it modelled a delivery system involving just eighty specially designed planes that could deliver more than a million tones of sulfuric acid, at a cost of approximately $2 billion a year. This is a minuscule amount of money compared with the expense of converting the world to a low carbon economy, or dealing with the costs of unfettered climate change.

The Calgary study also pointed out all the risks associated with such a strategy. More sulphur in the atmosphere means more damage to the ozone layer. It would also reduce the amount of rainfall and evaporation, so droughts, desertification and chronic water shortages would become common, with effects that might be worse than those of global warming. The sulphur haze would also produce a variable effect across different regions, creating winners and losers. And those are just the consequences thrown up by the study. Who knows what others risks planetary intervention on that scale might pose?

David Keith, the head of Calgary University's Institute for Sustainable Energy, Environment and Economy and the study's leader, also stressed the moral hazard attached to the project. There are two main risks: what he called "facile cheerleading" for solar management unaccompanied by serious efforts to reduce carbon emissions, and the question of how to govern a multilateral programme that would leave the losers vulnerable to environmental destruction.

Part of the moral difficulty with SRM techniques stems from their very affordability. $2 billion is well within the reach of an individual nation state, and it's possible to imagine a nation taking matters into its own hands and pressing ahead. But does any single country have the right to decide to implement a (hugely hazardous) programme that would affect the entire world's population in the most profound way? In fact, $2 billion is well within the reach of numerous extremely wealthy individuals, at which point possible geo-interventions start to look even more dystopian.

Then there's the fact that, once started, the programme would have a rapid cooling effect that would need to be constantly maintained for the rest of time. We all know that human beings tend not to be too good at planning anything much beyond the next five to ten years, fifty at push if you've got a genuinely determined (for which, read totalitarian) regime involved. Think of the terrorist threat to a mechanism that would, after some years of operation, plunge the Earth into an inhabitable state if it were "turned off." Even without terrorism, the entity in control of that mechanism might be very tempted to use it for military or economic gain. The legal challenges from the losers could tie the world up in knots. The consequences for the citizens of those loser regions, of course, would be hellish indeed.

There are no easy answers when it comes to geoengineering, because there are no easy solutions to climate change. The scale of the challenge is mind-curdling. But inaction is a morally feeble response. We need lots of serious, well-funded, long-term research, and we need it now. As Professor David Keith put it when he summed up his study, "Responsible management of climate risks requires deep emission cuts and research and assessment of SRM technologies. The two are not in opposition. We are currently doing neither; action is urgently needed on both."

60

FAILING GRACEFULLY, OR WHY EVERYTHING IS BETA NOW

Compared to the slow-motion disaster movie that is climate change, crashed sites and sub-optimal apps are problems that we can generally take in our stride—provided we've had our daily caffeine fix and aren't battling a deadline. In fact, there is no such thing as a finished digital product, and the most highly regarded applications (by those in the know) are not those that never fail, but those that fail gracefully. Where previously we wanted perfection from the things and services we consumed, now, as we grow used to living in a world where iterative design and Moore's Law dictate that everything is a work in progress, we are increasingly comfortable with the provisional, provided it serves its purpose. That's especially true when, in return for using a digital product at an earlier stage of its development (whether it's a game or a technical manual) we are asked to contribute our expertise or opinion to the work in progress.

Online everything is beta because the state of perfection is permanently receding on endless waves of innovation. And an app that is adaptable, or that can deliver a soft landing even when it fails, is far more valuable than the perfect-for-a-moment app that either lacks the flexibility to cope with whatever is coming down the line next, or is late. Good-Enough Right Now beats Very-Good Later, and completely defeats Was Perfect Once.

As we engineer more and more complex systems from vast amounts of code, we are developing our understanding that, with so many inputs, a consistently optimum outcome is simply impossible. The digital mindset is one that accepts that, in a perfect world, a new application would be perfection itself, but in reality it'll never be better than merely very good. This capacity to be very good, even in non-perfect conditions, does not happen by accident. It has been designed into the app, using the principle of failing gracefully as a guiding light.

To fail gracefully is to circumvent a crash. Failing gracefully is what occurs when, for example, a website built with a brand-new coding technique is encountered by an old browser that doesn't have the necessary capabilities. No, the browser will not display all the elements of the site, but it will not react by falling apart in a hissy fit and crashing; correctly designed, it will cope serenely, to the best of its ability, because it has been designed to be flexible. These are the apps beloved of coders everywhere; the apps that make even their failures look like successes, and the browsers that do the opposite, amplifying the mistakes in a webpage are, of course, much hated. A clever web designer too, will ensure that their design itself fails gracefully. Access a series of webpages made with Flash using the browser on the iPad, which has no Flash support, and you can see varying degrees of success at attempts to create designs that still work with the Flash content—pages that fail gracefully.

Failing gracefully is underpinned by a concept that comes as close as damn it to being a defining principle of Internet design.

The maxim "Be liberal in what you accept and conservative in what you send" was coined by Jon Postel, a legendary Internet engineer, but of course he merely put into words what the thousands of architects of the Internet put into the network, and the software that runs on it. Postel argued that the ideal to aspire to was, for example, an email programme that could accept any email, however broken, however corrupt the code, however out of date, and work with it successfully enough to display the message. The emails it generated itself, on the other hand, should be as near to flawless as possible, and in fact it should be working to fix any sub-standard emails received before it sent them on. Our Platonic ideal of an email programme would not crash if someone else had made a mistake and would never do anything to cause anyone else to crash. The ability to fail gracefully is a very noble quality.

Some products and situations lend themselves better to failing gracefully than others. A flawed retail website is one thing, a glitch in a council's website for paying taxes is quite another. Where money or personal safety correlate with digital complexity, even the most exquisitely designed app may not feel trustworthy enough. We have already seen that the financial industries have created a Singularity of complexity with their software, one that is incapable of failing gracefully on a consistent basis. There are other digital products in development that, though they sound exciting, are treated with scepticism by people who know a lot about software design.

Take the self-driving car, for example. It sounds like one of those Messianic technologies we looked at alongside nanotech, and may very well turn out to be. Google is at the forefront of the development of an autonomous car, though numerous vehicle manufacturers are also working on the concept. Its exponents claim that mass take-up would slash the number of deaths on the roads, once the pesky fallible humans have been removed from the equation. You wouldn't have to go far to find plenty of software engineers who would raise their eyebrows at this. It's tempting to imagine a safer road network with fewer poor drivers, but failing gracefully is not a concept that translates easily to a car with no driver, and especially one where you've been tempted to remove the pointless steering wheel. The same reasoning goes to explain the social, if not technical reason behind not having flying cars now that we're living in the future. A flying-car failure would be anything but graceful.

Most of us balk at the potential for disaster suggested by failing technology in such an obviously life-and-death situation, but we already live in a world where countless lives and limitless billions of dollars are dependent on the soft landings engineered by technology workers. And on a more everyday scale, we are evolving away from a natural philosophy of broken versus fixed, or in-progress versus finished. Even ten years ago, a new programme would go through closed beta testing in which a small group of selected people, colleagues or friends, would test a new app for flaws and bugs before it went live. These days beta tests are more often open affairs, involving hundreds if not thousands of volunteers. These people sign up to play a game, use a web app, or even read the first draft of a new manual on a programming language, and send their comments and criticisms back to the developers, writers, or publishers. There might be some risks or frustrations attached to using a product that's essentially still slightly broken, but the users gain access to the latest information or entertainment, and the glow of knowing that they are participating in collaborative work on something that has value to them. And why not: after all, the very good, though it never quite catches up with perfection, keeps on getting better.

61

WHY INFORMATION OVERLOAD IS ALL IN THE MIND

It is a truism of our age that we are bombarded with more information than we can handle. This makes us stressed and unhappy. We are helpless in the face of the onslaught; in extreme cases, little better than addicts whose screen habit is rotting our minds and destroying our concentration spans. We like to tell ourselves that our society has progressed too quickly for the health of our minds, that 500 years ago we were still an agrarian economy, living a less complex, usually rural existence and now most of us sit in front of computers at work and in our leisure time, hypnotized by the amount of information we are required to engage with. It's my contention that this is largely self-indulgent twaddle.

It's worth acknowledging that yes, the average office worker receives more than fifty emails a day (or eighty-five or a hundred—nobody really seems quite sure) and that we do indeed have access to and contact with an almost infinite amount of information. And of course it is true that some people feel compelled to pore over Facebook, and address the world via Twitter every ten minutes. The immediacy of social-networking platforms offers a compelling mechanism for people who are inclined to fret that they are missing out on fun, or who have a compulsive streak. If you are one of the increasing number of people who work for yourself, you might feel justified in checking your emails every half an hour: you believe you need to be quick off the mark to respond to an offer of work. So there is some justification to our anxious attachment to our screens—now that they provide access to the platform where we do more and more of our living, it's hardly surprising. Nonetheless, the idea that we are subject to more information than ever before, and furthermore that it is bad for us, and beyond our control, is flawed for several reasons.

The first is that we are kidding ourselves when we imagine that back in 1712 our ancestors lived a blissfully info-lite existence. They were in fact required to pay attention to all sorts of demands on their time and attention, the difference is that many of those demands did not arrive in a textual format that has been preserved. We live in an intensely text-centric world, where every memo or email has the potential to divert our attention from where it needs to be. But an eighteenth-century farmer or tailor or cook's maid had quite a lot of competing claims on their attention as well. And a great deal of the information they had to process derived from skill sets now lost to us, such as the ability to read the weather and the landscape if you were a farmer. That's before you factor in the information relayed through the oral culture of the time, plus authorities such as the Church and municipal powers. Our print-centric egos compel us to believe that no one has ever had it as overwhelming as we have, but that's simply not the case.

The second thing is that the hours devoted to Facebook and Twitter and blogs and YouTube are, obviously, completely optional. It does us no favours to tell ourselves that we cannot control how we choose to spend our leisure time. But even with the work-related email or research, or whatever it is that feels so unavoidable, it can always wait till the morning. We flatter ourselves if we think that we absolutely must remain connected at all times. Even if you're Chancellor Angela Merkel, you don't have to check your email every half an hour from rising until you retire to bed. If there's something really important, someone will call you. How much more does this apply to the rest of us?

Then there's the fact that, with the technological ability we now possess to filter information as it arrives, we can manage it far better than ever before, so that we see only what we need to. Arguably there's a bigger problem with not seeing things we should, given everything we know about the echo chamber.

Nevertheless, even if the scare stories are exaggerated—nobody has the concentration span to read Proust any more!—people do feel anxious. I firmly believe that, while this is understandable, it needs to be resisted, by reminding ourselves that we are in control and can learn to manage information with relatively minimal effort.

There are numerous ways to do this. One is to adopt a sort of Amish-inspired approach to the acquisition of the technological hardware. You might say, I only need a mobile phone that allows me to make calls and send text messages. The smartphone functions are not of interest to me. Or, I use my computer for word-processing only, so it doesn't matter that it's a 1993 model. This is of course a lifestyle choice and should not be confused with the sort of techno-poverty that developing countries are keen to lift their citizens out of.

An alternative is the digital sabbatical, in which you unplug from the Internet for a set period of time, anything from a few hours to a month. This is rooted in a common-sense grasp of the fact that if we want less information, we need to choose to connect less frequently to the source. My concern about the digital sabbatical, especially the extended kind (unless you're on holiday, in which case, carry on) is that it seems to me to encourage us to remain stuck in an anxiety-driven extreme. The feast or famine approach to the information riches of the Internet is not healthy. It's as if we'd arrived at the equivalent of the unlimited breakfast buffet and just couldn't stop eating the bacon. We need to learn moderation, not to resort to starving ourselves for a month.

Fortunately, we're seeing more and more moderation. With experience, people become more skilled and confident at choosing their Internet-based activities. Then there are features such as Gmail's priority inbox, where it learns to channel the things that are of most interest to you, and all sorts of timed Internet-use blockers. Many offices now operate a no-email Friday policy. If you want to communicate with a colleague, you have to talk to them; if you need to contact a supplier, you call them. Some companies are even taking out their intranets to reduce the amount of low-priority information buzzing around. In the last couple of years it has become more and more socially acceptable to check your email twice a day. In fact it is now the new orthodoxy of personal productivity. We will get better and better at living comfortably in our networked world. Both the technology and our evolving social contract will enable us to control the flow of information, not the other way round. Which is good news for fans of Proust.

62

THE LONG NOW

There is a paradox about the way we experience the Internet. On the one hand we might feel disorientated by the speed at which it has revolutionized, say, communication and retail; on the other, we can hardly remember a time before email and Amazon. This is a by-product of the extent to which the digital platform has suffused our everyday experience (and of human beings' ability to stop noticing what was once remarkable). Actually, the businesses and social networks and all the digital fallout that we have been examining are very new arrivals. Email has been a central part of the way we live for less than twenty years. Amazon, the closest thing on the Internet to an *ancien régime*, was founded in 1994. The pioneers who built the networks that enabled everything that came afterwards had a long-term vision—if not quite a plan—for what might be achieved, but even they admit to being astounded both by the pace of change and the way that change has been normalized.

The digital industries in the late Nineties and early Noughties were not always seen as template-changing early iterations of a phenomenon that was here to stay, even by their backers. In fact, there was a distinct feeling of disposability about them. The dot.com bubble was the latest in a centuries-long line of wild ventures that promised quick ways to make a fortune. For a while, the classic Silicon Valley start-up was basically a company designed to take full advantage of a short-term fad, a company that investors would cash out in a couple of years. A lot of those start-ups went bust; a few got sold on to corporations and made their founders and backers extremely wealthy before floundering almost immediately. (Meanwhile Amazon just grew and grew.) So there was a moment after the dot.com bubble burst in 2000 when, if you were a casual Internet user, you might have been forgiven for wondering whether it was really going to amount to much. After all, it was clear that very few of the first generation of Internet experimenters and entrepreneurs were thinking long-term.

But then, why would they? Nobody else was either. It took the stock-market crash of October 2008 and its ongoing catastrophic consequences to even raise a question about the doctrine of immediate results and permanent growth. After the haphazard scramble for bailouts and quick fixes, there is now a growing emphasis on financial and emotional investment in what has been termed "deep time" or the Long Now. There is an unprecedented level of interest in businesses and projects that are built to withstand shocks, that might grow incrementally, sustainably. Suddenly it doesn't look so clever to flip a start-up in eighteen months and get rich quick; in fact it looks a tiny bit distasteful compared to the alternative of growing a business that might last long enough to support a few other people too; even to make your kids proud of you.

It wasn't only in the digital sphere that the moment seemed ripe for investigating longer-term solutions to some of the global problems that were getting harder and harder to ignore. Even before the financial system's teetering turned into toppling there was an energy crisis waiting in the wings, not to mention looming environmental issues. If sustainability was to be achieved in any of these important areas, then everybody would need to get much more comfortable with making plans for the next fifty or one hundred years rather than five or ten.

That is, if we could just make it through the next few months. Modelling the management of disastrous changes of fortune, a standard exercise routinely undertaken by governments, has assumed a far greater priority since 2008. The scale and the unexpectedness of the credit crunch in the United States and Europe removed a great deal of complacency among elites (though many people would suggest, not quite enough). Nowadays the attitude is far more likely to be that anything could happen, rather than that it won't. But although it is diverting for the U.K.'s rulers to know precisely how long it would take for the country to descend to martial law in the event of, say, the euro's collapse coinciding with a dramatic rise in the price of oil triggered by conflict with Iran (answer: about two weeks), it is ultimately self-serving without a corresponding level of interest paid to fifty and hundred-year planning.

Elected politicians have rarely been the people to turn to if you want a long-term plan for getting anything done. Their habitual tendency not to think beyond their second term is exacerbated by the plethora of challenges and crises demanding their attention. It's very noticeable that the Chinese administration, unencumbered by either the need to win democratic elections or the possible collapse of the foundations of their society, are excelling at long-term planning. Chinese investment in South America and Africa is planned over a minimum of fifty years, to build the infrastructure and consumer confidence, not to mention the diplomatic ties, with potential trading partners and political allies.

Back in the developed North, systematic long-term plans of a political nature are generated by civil groups such as the Transition Town movement, which has so far attracted virtually no interest from central government of any stripe. The movement aims to build communities' resilience to the threats posed by peak oil, climate destruction and financial instability and works with local councils and other organisations to reduce energy use, food miles and waste. Started in 2005 by Rob Hopkins, a permaculture lecturer, it now has more than 400 transition initiatives in eight countries. Many of the transition movement's methods are extremely low-tech, but their communications networks take full advantage of Internet-based community building. Its fundamental principle—that if we plan for it properly, life after the encroaching crises will be better, not worse, than the way we live now—is determinedly hopeful, and pragmatic, about our long-term future.

It's difficult to know whether the U.S. military's Hundred-Year Ships project is a classic fantasy of salvation technology, or an instance of a dominant culture seriously trying to plan for its own demise. Quite probably it's both. This thought experiment models a hypothetical mission to a nearby solar system. Hundreds of people set off to establish a colony on an alternative world. Even if the ship were travelling at nearly the speed of light, it would be the grandchildren and great-grandchildren of the original crew members who arrived at their destination. Is it possible to plan a closed ecosystem capable of sustaining a community and preserving its social and cultural traditions in isolation, while minimising the risk of mutiny or civil war?

Even longer term than the Hundred-Year Ships are the projects of the Long Now Foundation. This philanthropic organisation was established in California in 1996 to foster long-term thinking and responsibility, over a timeframe of 10,000 years. Like the Slow Food movement, its purpose is to resist an accelerating culture, but in this case by projecting our interest way beyond the lifetimes of even our most distant descendents. Its most iconic plans are perhaps the Rosetta Project, a publically accessible digital library of the grammatical structures of all the world's languages, and the 10,000-year clock.

This magnificent mechanism is designed to be an inspiring symbol of deep time and a counterpoint to short-term thinking. The second prototype is being built and a mountain site in Nevada has been acquired for eventual installation, but as one might imagine no one is rushing to hit a deadline for completion. Work is steady, with new patents constantly emerging from intermediary stages. When it is finished the clock will, in the words of its designer Danny Hillis, "tick once a year, bong once a century and the cuckoo will emerge every millennium."

Hillis knows a thing or two about design: he's a legendary engineer, responsible for projects from super-computers to Disney theme-park rides. He thinks humanity needs his symbol of the Long Now to help us all live more responsibly in our infinitely brief present. And he might just be right.

63

JUST ENOUGH DIGITAL

When we discover something new that's fun and useful—or even just fun—we tend to overindulge in it. A new project occupies more of our time and attention than the old pastimes that have been tarnished by familiarity. The appeal of novelty is a truism that applies a hundred-fold to the Internet-based activities we've been examining. The Internet offers numerous rewards instantly, and more and deeper rewards as our usage becomes more sophisticated. It is entirely typical for someone to start using the Internet because they want to have an infinite source of recipes always at hand, only to find themselves investigating social networks, joining communities of interest, and writing a blog about their passion for recreating recipes from the Russian Imperial court. (Admittedly, that last detail may not in fact be typical, but the drive towards sophisticated Internet usage certainly is. It's fuelled by a rising sense of excitement—a "look what I can do with this!" realization.)

This is the journey that we have all been on, to a greater or lesser degree, over the fifteen or so years since the World Wide Web became part of daily life for people in the developed world. Everybody's existence has been altered. It has all been tremendously exciting and addictive for all of us. But actually, fifteen years is not a very long time. Not when we're talking about the effects of changes as far-reaching as these. We've been conducting a group experiment in ways to use this technology and we are only now starting to see an emergent best practice. What we think of as the new norms for business, for work / life balance, for creativity and politics and all the other myriad things we've been examining, might turn out in the long run to be symptomatic of a generalized fascination with extreme novelty.

This is not to say that the impact of the Internet-based technologies is overblown or that the Internet will turn out to be a transitory thing. It's not; it won't. It's the opposite: it changes everything and it's here to stay. But precisely because of that revolutionary power, it will take us a while to figure out how to use it for optimum effect.

There are trends emerging among long-term heavy users, particularly in areas such as email, social networking and other everyday tools. The pioneers who built the architecture and apps that have brought about the digital revolution are growing disenchanted with some aspects of the way the Internet has developed as it has torn through every sector of our lives. These are the same people who, as we saw in the discussion of the return to craft, are increasingly using their skills to make things that can be appreciated by non-programmers. They were the most excited of all of us, but are now suffering from queasiness. Overdosing on kitten videos can leave a nasty taste in the mouth. Now they want to move away from quantity (in terms of volume of material, time online) and towards better quality, in their own work and others'. They are developing a more refined palette, questioning the assertion that anything digital must be superior and that more is better, embracing techniques and software that moderate

Internet usage. This more skilful negotiation of the twenty-four-hour-a-day theme park of gratification that is the Internet only comes with experience. The kind of experience that results from immersion in the fifteen-year-long mass experiment.

There is accelerating disillusionment with the idea of hyper-connectedness, particularly in a business context. More and more people are rejecting high-church corporatism: the Blackberry buzzing on the bedside table. Permanent connection was an aspect of Internet-enabled behavior that we thought was useful, that our bosses frequently told us was essential. But the fact that it is technologically possible does not necessarily mean it's a good idea, at least on a social or individual psychological level. Since, as we've seen over and over again, etiquette around usage lags way behind the technological capabilities, it has taken us about ten years and a lot of stress to work out that, actually, you can't respond to every email you receive within twenty-four hours. Our Internet usage is going through a process of iterative design, motivated by that queasiness after over-indulgence and facilitated by the technology itself.

In the twenty-three years since the World Wide Web's inception there have been endless zero-sum game statements made by detractors and enthusiasts alike. The Internet will destroy culture, brainwash our children, rot our ability to concentrate, think for ourselves etc. etc. Or, conversely, in a shrinking world we can all be connected, productivity of every kind will soar; in cyberspace we can all be creatives, freed of any need for talent or skill by the magic technology. In fact, as we are now realizing, almost all such pronouncements are unhelpful and misleading. The early adopters, who never paid much attention to the Internet's detractors, are increasingly resistant to uncritical adulation. The Internet doesn't rot anyone's brain, but eight hours of kitten videos a day might. Being able to watch your baby nephew take his first steps on the other side of the world is good; being connected every moment you are awake is not.

There is every reason to be thrilled by the networked world we now live in, but there is no need to abandon either our critical faculties when using its apps, or our sense that we have the power to optimize its place in our lives. If we are now revising our initial boundless enthusiasm, that does not imply that we are turning our backs on all the glorious things we've discovered. Call it optimization in computing terms, or plain old growing up in human development terms—either way, we are moving towards just enough digital.

64

THE ZEN OF DIGITAL LIVING

In the mid-1990s the old media was awash with predictions that this Internet thing would turn out to be a fad. They were followed by wave after wave of assertions that the Internet was in fact here to stay but was incredibly corrosive of all we held dear and needed to be resisted (though quite how, when it was taking over the world, was never really clear).

There is nothing new about people being alarmed by technological innovation of course, but what is different in the case of the Internet is the extent to which this particular technological innovation has revolutionized every aspect of life, and with it, brought excitement and anxiety on an unprecedented scale. In 1999 Douglas Adams wrote an article for the *Sunday Times* entitled "How to Stop Worrying and Learn to Love the Internet" that set out the three stages of human beings' attitude to any new technology, but most particularly to this one. Adams' take on our default position is that "anything that's already in the world when you're born is just normal." That's followed by "anything that gets invented between then and before you turn thirty is incredibly exciting and creative and with any luck you can make a career out of it," and then, inevitably, "anything that gets invented after you're thirty is against the natural order of things and the beginning of the end of civilization as we know it until it's been around for about ten years when it gradually turns out to be all right really."

The most important element of Adams' pronouncement is that first one, about normality. For many people, Internet-based applications like Skype and email barely register as technology any more. They are just features of the way the world works, like cold in winter and heat in summer. We all of us have the capacity to take for granted the previously miraculous innovations that excited or worried previous generations, but we are all on a conveyor belt towards excitement and worry in our turn. One of the great challenges of the immediate future will be to strengthen a flow of compassion between Adams' three camps, whose world views are so schismatically different. It's imperative that we do. The consequences of a failure in the dialogue between, say, the legislators that were way past thirty when the World Wide Web became a phenomenon, and the latest generation of activists who cannot imagine a world without Facebook and file-sharing, are worrying. Any attempt to roll back Internet freedom by the digitally illiterate would be a terrible waste of time, effort and money. The frustration of the younger digital natives would spill over into protest and even more disenchantment with the hierarchies that rule them.

Adams' scheme accounts for a drift towards acceptance, as fear subsides and the new technology becomes part of the furniture. To a great extent this has happened with the Internet, but the digital world still provokes fear and anger, perhaps because of its all-pervasiveness and its ability to disrupt entrenched power systems on a massive scale. It behoves those who live in the Internet-enabled world with most ease to extend empathy to those who are still struggling to grasp what has happened. That may prove difficult. Across the developed world there is already an inter-generational conflict raging over the monopoly of assets and power by the baby boomers even as the privileges they enjoyed are disappearing from their children's horizons. With the global economy in a parlous state, the potential for this fight to escalate is huge—and if it does, it will be played out over the Internet.

Relations between those who do get it and those who don't are one thing. How to interact with the digital world is an ongoing challenge for the digital natives as well. The first and second generation have moved from excited first contact with the basics of a networked environment, via exhilarating immersion in its complexity, to, increasingly, a pared-back simplicity in their Internet usage that allows them to live harmoniously with and on the new platform. These people have been living online for nearly twenty years and during that time they have been evolving the sort of best practice that we saw in the previous chapter, increasingly unsubscribing to information, connecting only when it suits them and learning how to absorb the extraordinary capabilities imparted by the new technology into their own range of skills.

Over the last ten years, improvements in personal communications technology have been nothing short of transformative. It is almost as if we have all acquired superpowers—the ability to open up a live window into the experience of someone on the other side of the world, for example, via Skype. We can store limitless email exchanges and recall them at the touch of a button to check what we really said to that acquaintance we fell out with six years ago. There's no need to remember friends' birthdays or which actor starred in which film—if we need to know the date or the name we can look it up in a split second. We need never get lost again since so long as we have a smartphone we can always locate ourselves, virtually anywhere in the world.

These properties belong to the technology, of course, not to us. And yet, because of the ever-increasing ubiquity and intimacy of our usage of these devices, we absorb their powers in the same way that if we lived on the moon we would all be able to jump ten feet in the air with no effort at all.

The memory in our smartphones or our personal computers is increasingly an extension of our own selves since we have outsourced so much of our cognitive function to these technologies. And as they grow ever more widespread and ever more personal, we begin to assume that everyone else around us is also using the same technology; that they too are taking on these superpowers. In the age of GPS apps on every handheld device, it is less and less comprehensible to excuse our lateness to an appointment on the grounds of getting lost. We have seen over and over again that the etiquette that governs our use of the new capabilities evolves far more slowly than the technology itself. The digital natives have spent the last ten years developing a consensus about acceptable behavior between individuals in the new digital world, and just recently those norms have taken a distinct shift towards a relaxed simplicity. Now as boundaries between our own consciousness and the technology's functions grow more and more porous, we need to set to work again, to develop a new sense of where the limits are.

In the end, the greatest challenge of digital living will be to manage the relationship we have with our own selves, as even that most intimate and nebulous connection is more and more mediated by digital technology. It is early days, but its glimmerings can be seen in the emergence of the precepts of personal productivity gurus. These figures, twenty-first-century prophets promising peace of mind and perfect mastery of your to-do list, encourage us, for example, to aim for the state of inbox zero. When our email is an ever-present record of our social and work interactions, a list of obligations and incomplete tasks all rolled into one, it begins to feel like an extension of our unquiet mind.

Personal productivity systems teach us to inspect each item in the email system (and the box file and the document files on our computer) consider its purpose in a calm and mindful way and resolve to move it on or out of our lives. This is a mind-clearing technique for our times, a humble successor to the ritual of confession or therapy, but an analogous tool. Perhaps, before too long, an empty inbox and a clear conscience will be one and the same thing.

One thing is for sure, the Internet is not going away. It is the essence of the world we live in, the dominant paradigm for all social cultural and economic interactions in the twenty-first century. It's a staggering achievement, a tool for the reinvention of society, a giant experiment in new ways to relate to people, do business and learn about the world; it's also an infinitely capacious junk store bursting with videos of kittens, a glorious joke that's evolving its own complexity, one that may yet outwit us. The Internet has shaped us and will continue to define the contours of our endeavours for the foreseeable future. It is us and we are it.

INDEX

Printed in the United States
by Baker & Taylor Publisher Services